# Keplerian Ellipses
# (Second Edition)

A student guide to the physics of the gravitational two-body problem

**Online at: https://doi.org/10.1088/2514-3433/acb430**

## About the program:

AAS-IOP Astronomy ebooks is the official book program of the American Astronomical Society (AAS) and aims to share in depth the most fascinating areas of astronomy, astrophysics, solar physics, and planetary science. The program includes publications in the following topics:

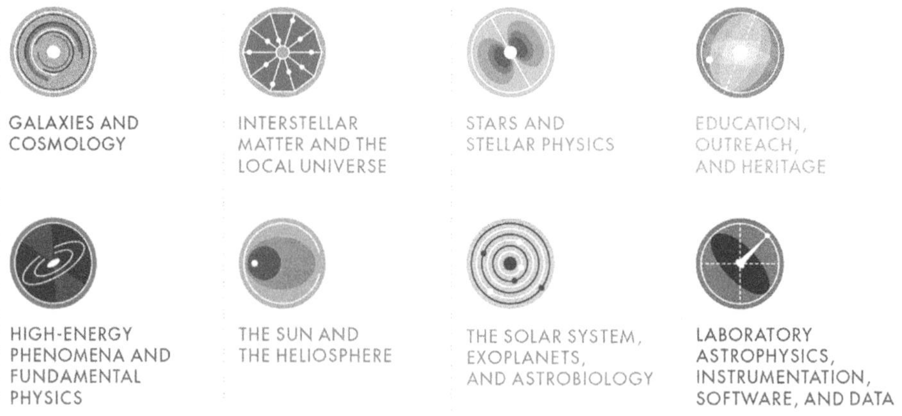


Books in the program range in level from short introductory texts on fast-moving areas, graduate and upper-level undergraduate textbooks, research monographs, and practical handbooks.

For a complete list of published and forthcoming titles, please visit iopscience.org/books/aas.

## About the American Astronomical Society

The American Astronomical Society (aas.org), established 1899, is the major organization of professional astronomers in North America. The membership (~7,000) also includes physicists, mathematicians, geologists, engineers, and others whose research interests lie within the broad spectrum of subjects now comprising the contemporary astronomical sciences. The mission of the Society is to enhance and share humanity's scientific understanding of the universe.

# Editorial Advisory Board

# Keplerian Ellipses (Second Edition)

A student guide to the physics of the gravitational two-body problem

**Bruce Cameron Reed**
*Alma College, Alma, MI, USA*

**IOP** Publishing, Bristol, UK

ISBN 978-0-7503-5608-4 (ebook)
ISBN 978-0-7503-5606-0 (print)
ISBN 978-0-7503-5609-1 (myPrint)
ISBN 978-0-7503-5607-7 (mobi)

DOI 10.1088/2514-3433/acb430

Version: 20230301

AAS–IOP Astronomy
ISSN 2514-3433 (online)
ISSN 2515-141X (print)

British Library Cataloguing-in-Publication Data: A catalogue record for this book is available from the British Library.

Published by IOP Publishing, wholly owned by The Institute of Physics, London

IOP Publishing, No.2 The Distillery, Glassfields, Avon Street, Bristol, BS2 0GR, UK

US Office: IOP Publishing, Inc., 190 North Independence Mall West, Suite 601, Philadelphia, PA 19106, USA

Cover image: Orbits of planets in the Solar System. Image credit: Mark Garlick / Science Photo Library.

# Contents

# Preface

Celestial mechanics is the crown jewel of Newtonian physics. Building on the observational and theoretical work of Galileo, Tycho, Copernicus, Kepler, and others, Isaac Newton deployed his hypothesized law of universal gravitation and his newly developed calculus to show why the orbits of the Moon and planets have the shapes that they do: Kepler's three laws of planetary motion; see Figures P.1 and P.2. This was a stunning development in human intellectual history. Contrived systems of equants, deferents, and epicycles gave way to a single elliptical orbit for each planet, dictated by a semimajor axis and eccentricity (Table P.1). This new approach facilitated predictions of celestial phenomena to unprecedented levels of accuracy, and the realization that these orbits reflected an underlying universal physical law whose validity extended from the firmament of the Earth to the remotest cosmos marked the birth of modern science.

This book presents a self-contained treatment of the Newton/Kepler two-body orbit problem at the level of an advanced undergraduate physics student.

Any student of physics or astronomy will surely encounter Kepler's three laws of planetary motion in a dynamics or astrophysics class. Proofs of these laws are some of the ultimate products of this book, so let's summarize them right away, expressed in modern parlance:

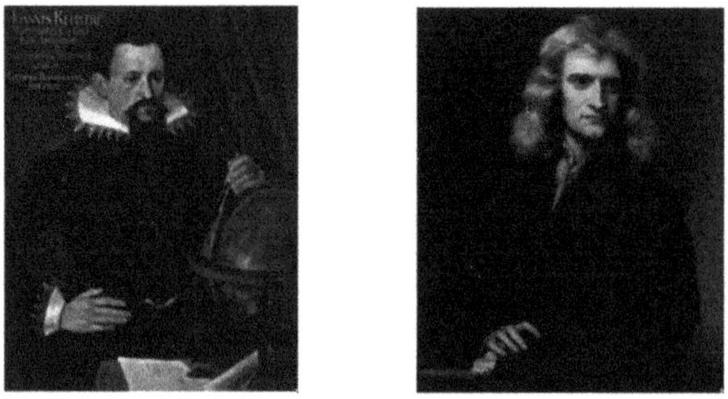

**Figure P.1.** Left: Johannes Kepler (1571–1630). Source: https://upload.wikimedia.org/wikipedia/commons/d/de/JKepler.png. Right: Sir Isaac Newton (1643–1727). Source: https://commons.wikimedia.org/wiki/Isaac_Newton#/media/File:Sir_Isaac_Newton_(1643-1727).jpg. These images [JKepler; Portrait of Sir Isaac Newton, 1689] have been obtained by the authors from the Wikimedia website, where they are stated to have been released into the public domain. It is included within this article on that basis.

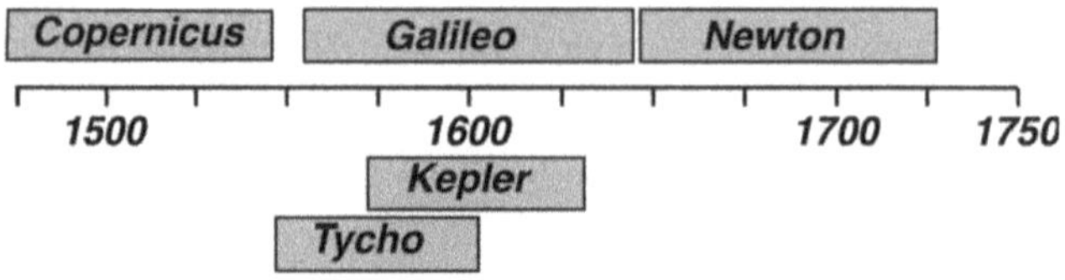


**Figure P.2.** Timeline of the lives of some of the founders of modern astronomy. Copernicus (1473–1543), Tycho (1546–1601), Galileo (1564–1642), Kepler (1571–1630), and Newton (1643–1727).

**Table P.1.** Solar-system Orbital Data.

| Object | Symbol | Semimajor Axis (AU) | Synodic period (days) | Sidereal period (yr) | Eccentricity | Mass (Earth = 1) |
|---|---|---|---|---|---|---|
| Sun | ☉ | ... | ... | ... | ... | 332,946 |
| Mercury | ☿ | 0.3871 | 115.9 | 0.2408 | 0.2056 | 0.0553 |
| Venus | ♀ | 0.7233 | 583.9 | 0.6152 | 0.0068 | 0.8150 |
| Earth | ⊕ | 1.0000 | ... | 1.0000 | 0.0167 | 1.0000 |
| Mars | ♂ | 1.5237 | 779.9 | 1.8808 | 0.0934 | 0.1075 |
| Jupiter | ♃ | 5.2029 | 398.9 | 11.863 | 0.0484 | 317.83 |
| Saturn | ♄ | 9.537 | 378.1 | 29.447 | 0.0539 | 95.159 |
| Uranus | ♂ | 19.189 | 369.7 | 84.02 | 0.0473 | 14.50 |
| Neptune | ♆ | 30.0699 | 367.5 | 164.79 | 0.0086 | 17.20 |
| Ceres | ... | 2.7658 | 466.6 | 4.600 | 0.078 | <0.001 |
| Pluto | ♇ | 39.482 | 366.7 | 248.021 | 0.2488 | 0.002 |
| Haumea | ... | 43.34 | ... | 285.4 | 0.189 | <0.001 |
| Makemake | ... | 45.79 | ... | 309.9 | 0.159 | <0.001 |
| Eris | ... | 67.67 | ... | 557.2 | 0.4418 | 0.003 |
| Sedna | ... | 525.6 | ... | ~12,050 | 0.855 | <0.001 |
| Halley's comet | ... | 17.834 | ... | 75.3 | 0.967 | <0.001 |

Ceres was traditionally termed an asteroid; it is now considered to be a dwarf planet. Pluto, Haumea, Makemake, Eris, and Sedna are "plutoid"-class dwarf planets, essentially round objects orbiting the Sun at distances greater than that of Neptune. Semimajor axes are in terms of Astronomical Units (AUs), the mean Earth/Sun distance, and periods are in years.

1AU = $1.495\,98 \times 10^{11}$ m.

One year = 365.25 days = 31,557,600 seconds = $3.156 \times 10^7$ seconds.

$M_{Earth} = 5.972 \times 10^{24}$ kg.

$M_{Sun} = 1.988 \times 10^{30}$ kg.

Radius of Earth = 6371 km.

Newtonian gravitational constant: $G = 6.674 \times 10^{-11}$ m$^3$ kg$^{-1}$ s$^{-2}$.

$GM_{Sun} = 1.327 \times 10^{20}$ m$^3$ s$^{-2}$

Data adopted from https://www.princeton.edu/willman/planetary_systems/Sol/ and various online sources. See also Table 5.1.

*Kepler's first law*: Each planet orbits the Sun on an elliptical path, with the Sun at one focus of the ellipse.

*Kepler's second law*: A line from the Sun to a planet sweeps out equal areas in equal times.

*Kepler's third law*: The square of a planet's orbital period is directly proportional to the cube of the semimajor axis of its elliptical orbit.

Kepler published his first two laws in his *Astronomia Nova* (New Astronomy) in 1609; these were deductions based on Tycho Brahe's detailed observations of the motions of Mars. The third law appeared in his 1619 *Harmonices Mundi* (The Harmony of the World).

Orbital mechanics has been picked apart for over 400 years (over 2000 if you count Ptolemy); the physics sections of school libraries are rife with dynamics texts, and extensive discussions of Keplerian mechanics can be found online. Why, then, did I prepare this book?

When students encounter orbital mechanics, it is usually as a unit of a much more extensive dynamics class that will include numerous topics such as resisted motion, rotating reference frames, damped and driven vibrational motions, Lagrangian and Hamiltonian dynamics, least-action principles, and perhaps even relativity. In such a vast docket of material to be covered in one or two semesters—all worthy and containing much powerful physics—planetary orbits becomes just one more topic to squeeze into the syllabus. It is difficult to do this beautiful topic justice in such circumstances. This book grew out of my conviction that students deserve, can benefit from, and would, I hope, enjoy a brief, affordable, self-contained treatment of the essential physics of Keplerian elliptical orbits. When you have finished reading this book I hope that you will keep it as a reference/refresher.

My goal has been to prepare a volume that starts with fundamental concepts such as position, velocity, acceleration, force, energy, momentum, and angular momentum expressed in polar coordinates, and then pair them with Newton's gravitational law to show you how elliptical orbits and Kepler's laws result, while also setting up very general expressions for exploring relationships between distances, angles, and times for such orbits. The derivations presented in this book are not complete in every detail, but are given in enough detail that you should be able to fill in the gaps. And do fill in the gaps: think of them as exercises in algebra practice which will help reinforce the concepts involved.

This book comprises eight chapters and four appendices. Since gravity is a "central" force, that is, one which acts directly along the line joining the centers-of-mass of two objects, polar coordinates are the natural framework within which to analyze orbital mechanics. Chapter 1 is devoted to a brief review of polar coordinates and the properties of their corresponding unit vectors, and also lists a few useful integrals and identities. Chapter 2 develops expressions for position, velocity, acceleration, angular momentum, torque, and energy in polar coordinates. Since these expressions do not assume any particular force law (that comes later), they are very general. Chapter 3 introduces the idea of a central force in polar-coordinate form; shows how the mutual interaction of two bodies under the action of a central force can be reduced to considering that of a single "reduced" mass imagined to be orbiting another mass which serves as the origin of the coordinate system; shows why angular momentum must be conserved within such a system; sets up very general integrals for expressing relationships between time, angle, and radial position for such motions; and shows how Newton's second law $\boldsymbol{F} = m\boldsymbol{a}$ can be transformed into an expression that relates rates of change of radial position to the angular position of the orbiting body as opposed to rates of change with respect to time. All of this can be done without yet specifying the particular (i.e., gravitational) form of the central force. It is also shown that there are only two possible types of central forces which satisfy "shell-point equivalency," that is, that a shell of material of some finite radius can be considered to be shrunk to a point mass located at its center so far as the force it exerts on an external object is concerned.

It is due to Johannes Kepler that we know that planets orbit the Sun in elliptical paths with the Sun at one focus of the ellipse. Chapter 4 reviews some of the properties of ellipses, and shows how the polar-form equation for an ellipse, which is so handy for analyzing orbits, can be set up from the idea of constructing an ellipse by considering the curve generated by a string of fixed length which wraps around two fixed "foci." This chapter also presents a proof of Kepler's second law by examining the area of an ellipse, and briefly describes how Kepler actually went about plotting planetary orbits and establishing their sidereal and synodic periods. This sets the stage for Chapter 5, where it is shown by two methods that ellipses satisfy the force equation set up in Chapter 3 in the case of Newton's law of gravitation and that Kepler's three laws result. This chapter also takes up some corollary issues, namely the Laplace–Runge–Lenz vector, which offers an alternate approach to Keplerian orbits, how Kepler's third law plays out for circular orbits involving non-inverse-square forces, the concept of effective potential, a brief foray into perturbation theory, and the concept of escape velocity. Chapter 6 is devoted to a derivation of "Kepler's equation," a compact although not analytically solvable expression which relates the position of a planet in its orbit to time. Chapter 7 examines how spacecraft can be put into transfer orbits to have them rendezvous with another planet or target object such as an incoming asteroid. Chapter 8 considers a number of miscellaneous issues such as determining the average distance of a planet from the Sun, the average speed of a planet in its orbit, why the *James Webb Space Telescope* was launched to a location so far from the Earth, the famous perihelion advance of Mercury, an approach to unit conversions, the orientation of Earth's own orbit, gravitational "scattering," and the motion of the Sun about the center of mass of the solar system. The appendices provide a summary of spherical coordinates, a deeper look at perturbation theory for circular orbits, a brief bibliography for readers who wish to pursue more detailed study, a summary of key formulae, and a glossary of the many mathematical symbols used throughout the text.

# Acknowledgments

My interest in orbital mechanics was stimulated in my own student days by a number of excellent teachers at both the University of Waterloo and Queen's University, notably Drs. Pim FitzGerald, Richard Henriksen, James Leslie, Donald Rayburn, and Robert Snyder. A number of instructors, students, friends, department colleagues, and family members have supported and encouraged me throughout my career. It gives me pleasure to especially acknowledge Karen Ball, Dick Bowker, Peter Burns, Peter Dawson, Michael DeRobertis, Eugene Deci, Carleen Dewit, Patrick Furlong, John Gibson, Bob Hayward, Lorraine Hill, Lisa Jylänne, Patricia Kinnee, Vern Koslowsky, Gilles Labrie, Pat Mueller, Lorne Nelson, John Palimaka, John Schreiner, and Ute Stargardt. Sadly, some of these wonderful people are no longer among us but are fondly remembered: *Requiesce in pace.*

This work, like others which have preceded it, is once again dedicated to my wife Laurie, who has continued to bear with my endless mutterings about arcane integrals and general distractedness. Our various cats over the years, Fred, Leo, Stella, Cassie, Nyx, and Newton, have never failed to remind me who is in charge.

*Johannes Kepler set out to discover India and found America. It is an event repeated over and again in the quest for knowledge. But the result is indifferent to the motive. A fact once discovered, leads to an existence of its own, and enters into relations with other facts of which their discoverers never dreamt. Apollonius of Perga discovered the laws of the useless curves which emerge when a plane intersects a cone at various angles: these curves proved, centuries later, to represent the paths followed by planets, comets, rockets, and satellites.* [Arthur Koestler, The Sleepwalkers (1959); reprinted by Grosset & Dunlap, New York, 1963, p. 398.]

*I do not know what I may seem to the world, but, as to myself, I seem to have been only like a boy playing on the sea shore, and diverting myself in now and then finding a smoother pebble or a prettier shell than ordinary, while the great ocean of truth lay all undiscovered before me.* [Attributed to Newton not long before his death; in Richard S. Westfall, Never at Rest: A Biography of Isaac Newton (1980) Cambridge University Press, p. 863.]

# Author biography

## Bruce Cameron Reed

Bruce Cameron Reed is the Charles A. Dana Professor of Physics Emeritus at Alma College, in Alma, Michigan. He holds a PhD in Physics from the University of Waterloo in Canada. In addition to a quantum mechanics text and five books on the Manhattan Project (including the IOP Concise Physics volumes *Atomic Bomb: The Story of the Manhattan Project* and *The Manhattan Project: A Very Brief Introduction to the Physics of Nuclear Weapons*), he has published over 150 papers in peer-reviewed scientific journals on research in the fields of astronomy, data analysis, quantum physics, mathematics, nuclear physics, the history of physics, and the physics of nuclear weapons. In 2009 he was elected a Fellow of the American Physical Society "For his contributions to the history of both the physics and the development of nuclear weapons in the Manhattan Project." After a 35-year teaching career spent in both Canada and the United States, he is now semiretired and living in Nova Scotia with his wife Laurie and a variable number of cats. He continues to teach part-time, pursue various publication projects, and serves as an Associate Editor with *American Journal of Physics*.

# Chapter 1

# Polar Coordinates—A Review

## 1.1 Fundamental Definitions

The system of polar coordinates adopted in this book is illustrated in Figure 1.1 and is standard in most mathematical physics texts: $r$ is the radial distance from the origin to the point of interest ($0 < r < \infty$), and $\phi$ is the "azimuthal" angle, measured counterclockwise from the positive-$x$-axis in the $xy$-plane ($0 \leqslant \phi \leqslant 2\pi$). I will also refer to $\phi$ more colloquially as the "polar" angle, although one must be careful with this as there is a different polar angle in spherical coordinates.

The transformation equations from Cartesian to polar coordinates are

$$\begin{aligned} r &= \sqrt{x^2 + y^2} \\ \phi &= \tan^{-1}(y/x), \end{aligned} \tag{1.1}$$

and the reverse transformations from polar to Cartesian coordinates are

$$\begin{aligned} x &= r \cos\phi \\ y &= r \sin\phi. \end{aligned} \tag{1.2}$$

## 1.2 Polar Coordinate Unit Vectors

We will make considerable use of polar coordinate unit vectors. Vectors will always be indicated by ***bold italic*** type, and unit vectors by bold italic type with a caret (^), such as $\hat{\boldsymbol{r}}$. In terms of the Cartesian unit vectors, the polar coordinate unit vectors are

$$\begin{aligned} \hat{\boldsymbol{r}} &= (\cos\phi)\hat{\boldsymbol{x}} + (\sin\phi)\hat{\boldsymbol{y}} \\ \hat{\boldsymbol{\phi}} &= -(\sin\phi)\hat{\boldsymbol{x}} + (\cos\phi)\hat{\boldsymbol{y}}. \end{aligned} \tag{1.3}$$

$\hat{\boldsymbol{r}}$ can be derived by inspection, as it is defined by $\hat{\boldsymbol{r}} = \boldsymbol{r}/r$. To derive $\hat{\boldsymbol{\phi}}$, imagine writing it as $\hat{\boldsymbol{\phi}} = \phi_x\hat{\boldsymbol{x}} + \phi_y\hat{\boldsymbol{y}}$, where $\phi_x$ and $\phi_y$ are components to be determined. Since $\hat{\boldsymbol{\phi}}$ is to be a

 

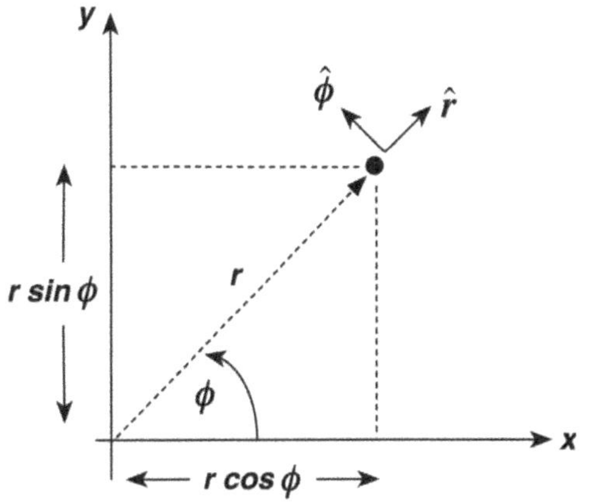


**Figure 1.1.** Polar coordinates $(r, \phi)$ as used in this book. The arrows attached to the point of interest (the black dot) show the directions of the polar unit vectors at that point.

unit vector, the components must be such that $\phi_x^2 + \phi_y^2 = 1$. Also, since $\hat{r}$ and $\hat{\phi}$ are to be perpendicular, their dot product must be zero: $\hat{r} \cdot \hat{\phi} = \cos\phi(\phi_x) + \sin\phi(\phi_y) = 0$. From these you can solve for $\phi_x$ and $\phi_y$. There will be two solutions; the convention is to have a unit vector point in the direction in which its corresponding coordinate increases; that is, we want $\hat{\phi}$ to point in the $\hat{y}$ direction when $\phi = 0$; the sign choice in Equation (1.3) makes this so.

Conversely, one can solve Equations (1.3) to express the Cartesian unit vectors in terms of the polar unit vectors as

$$\begin{aligned} \hat{x} &= (\cos\phi)\,\hat{r} - (\sin\phi)\,\hat{\phi} \\ \hat{y} &= (\sin\phi)\,\hat{r} + (\cos\phi)\,\hat{\phi}. \end{aligned} \tag{1.4}$$

*This is important: Cartesian unit vectors are considered to be fixed in space and act like constants, while polar unit vectors are functions of the direction $\phi$ of the point under consideration. Polar unit vectors do not act like constants.*

Like the Cartesian system, polar coordinates form a right-handed system, with the same rules for forming scalar ("dot") and vector ("cross") products. A useful result along this line is that $\hat{r} \times \hat{\phi} = \hat{z}$, where $\hat{z}$ is the usual Cartesian $z$-direction unit vector which emerges from the page, perpendicular to both $\hat{x}$ and $\hat{y}$. You can prove this by crossing the two expressions in Equation (1.3) or by applying the right-hand-rule to Figure 1.1. Figure 1.2 shows pictorial triads as a way of remembering how cross-products work. Suppose that you wish to compute $\hat{r} \times \hat{\phi}$ as described above. Start by locating $\hat{r}$ in the triad. Follow the circle until you get to $\hat{\phi}$ by the shortest route; this requires a counterclockwise motion. Keep going counterclockwise until you hit the next unit vector in the triad, which is $\hat{z}$. The convention is that a counterclockwise motion is taken to be positive, while clockwise is negative. Also as with Cartesian unit vectors, any polar unit vector dotted into itself gives unity, and crossed into itself gives zero.

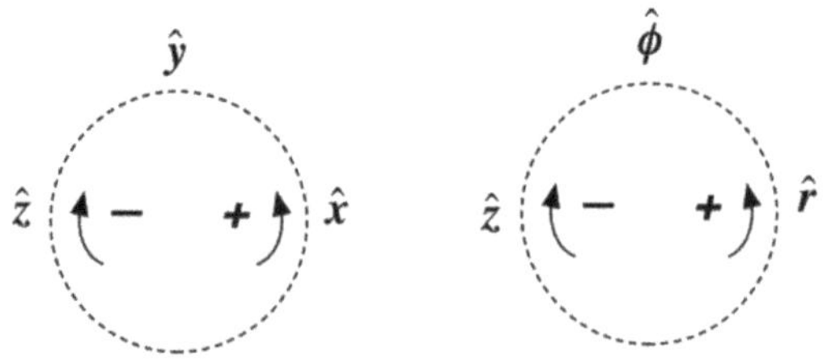


**Figure 1.2.** Mnemonic triads for remembering the results of unit-vector cross-products. If a counterclockwise movement is required to bring the first unit vector in a cross-product into the second one, the result is the positive of the next unit vector found by proceeding in the same direction. A clockwise movement gives the negative of the next unit vector found by proceeding in the same direction.

In general, the polar system is three-dimensional, with unit vectors ($\hat{\boldsymbol{r}}$, $\hat{\boldsymbol{\phi}}$, $\hat{\boldsymbol{z}}$), with any general vector $\boldsymbol{V}$ expressible in the form $\boldsymbol{V} = V_r\hat{\boldsymbol{r}} + V_\phi\hat{\boldsymbol{\phi}} + V_z\hat{\boldsymbol{z}}$, where any or all of the components ($V_r$, $V_\phi$, $V_z$) can be functions of any or all of the coordinates ($r$, $\phi$, $z$). Usually we will only be concerned with the $V_r$ and $V_\phi$ components, although the $z$-component does turn up in the formulation of angular momentum in the following chapter. It should be noted that some texts use the symbol $\rho$ in place of $r$, reserving the latter for use as an origin-to-point-of-interest distance in spherical coordinates. Since the physical application here is orbital motion in the $xy$-plane where it is traditional to use $r$ as a Sun-planet distance, I will stick to this convention; no confusion should arise.

## 1.3 Time Derivatives of Polar Coordinate Unit Vectors

For later calculations, it will be very handy to have expressions for the time-derivatives of the polar unit vectors in terms of themselves. That for $d\hat{\boldsymbol{r}}/dt$ is done here as an example. Begin by taking the time-derivative of the expression for $\hat{\boldsymbol{r}}$ in Equation (1.3), remembering that as the physical point of interest ($r$, $\phi$) moves, $\phi$ will be a function of time. Treating the Cartesian unit vectors as constants gives

$$\frac{d\hat{\boldsymbol{r}}}{dt} = \left[-\sin\phi\left(\frac{d\phi}{dt}\right)\right]\hat{\boldsymbol{x}} + \left[\cos\phi\left(\frac{d\phi}{dt}\right)\right]\hat{\boldsymbol{y}}. \tag{1.5}$$

As a notational convenience, it is customary to designate time-derivatives with an overlying dot, a notation developed by Newton himself. So, for example, $d\phi/dt$ is compacted to $\dot{\phi}$: be careful to keep an eye peeled for these dots. This is not done with the derivatives of unit vectors such as $d\hat{\boldsymbol{r}}/dt$, however, to avoid losing the dot in the caret. In this notation, Equation (1.5) becomes, on factoring out $(d\phi/dt) = \dot{\phi}$,

$$\frac{d\hat{\boldsymbol{r}}}{dt} = (\dot{\phi})[-\sin\phi\,\hat{\boldsymbol{x}} + \cos\phi\,\hat{\boldsymbol{y}}]. \tag{1.6}$$

The term inside the square brackets here is exactly $\hat{\boldsymbol{\phi}}$ from Equation (1.3):

$$\frac{d\hat{\boldsymbol{r}}}{dt} = (\dot{\phi})\,\hat{\boldsymbol{\phi}}. \tag{1.7}$$

The time-derivative $\hat{\boldsymbol{\phi}}$ follows similarly:

$$\frac{d\hat{\boldsymbol{\phi}}}{dt} = -(\dot{\phi})\,\hat{\boldsymbol{r}}. \tag{1.8}$$

These expressions will prove very valuable in the following chapter. In this book I will generally follow the convention of writing polar-coordinate vectors with components in the order $(\hat{\boldsymbol{r}}, \hat{\boldsymbol{\phi}})$.

**Exercise:** For practice, derive an expression for $(d\hat{\boldsymbol{r}}/dt) \times (d\hat{\boldsymbol{\phi}}/dt)$.

Answer: $(\dot{\phi})^2\hat{\boldsymbol{z}}$.

## 1.4 Some Useful Integrals and Expansions

The following definite integrals will be useful in later calculations. Various indefinite integrals will also arise, but these will be dealt with as needed. We will always have $0 \leqslant \varepsilon < 1$.

$$\int_0^{\pi} \frac{d\phi}{(1 - \varepsilon \cos \phi)} = \frac{\pi}{(1 - \varepsilon^2)^{1/2}}. \tag{1.9}$$

$$\int_0^{\pi} \frac{d\phi}{(1 - \varepsilon \cos \phi)^2} = \frac{\pi}{(1 - \varepsilon^2)^{3/2}}. \tag{1.10}$$

$$\int_0^{\pi} \frac{d\phi}{(1 - \varepsilon \cos \phi)^3} = \frac{\pi(1 + \varepsilon^2/2)}{(1 - \varepsilon^2)^{5/2}}. \tag{1.11}$$

$$\int_0^{\pi} \frac{\cos \phi \, d\phi}{(1 - \varepsilon \cos \phi)^2} = \frac{\pi\, \varepsilon}{(1 - \varepsilon^2)^{3/2}}. \tag{1.12}$$

$$\int \frac{dx}{(c + d \cos x)^2} = \frac{1}{(d^2 - c^2)}\left[\frac{d \sin x}{(c + d \cos x)} - c\int \frac{dx}{(c + d \cos x)}\right]. \tag{1.13}$$

Integral (1.10) is a special case of (1.13), with the solution of the integral within the square brackets depending on whether $c^2 > d^2$ or $d^2 > c^2$. Our concern will be with the former case, in which case one has

$$\int \frac{dx}{(c + d \cos x)} = \frac{2}{\sqrt{c^2 - d^2}} \tan^{-1}\left[\sqrt{\frac{c - d}{c + d}} \tan\left(\frac{x}{2}\right)\right]. \tag{1.14}$$

We will also encounter

$$\int \frac{dx}{x\sqrt{cx^2 + bx + d}} = \frac{1}{\sqrt{-d}} \sin^{-1}\left[\frac{bx + 2d}{|\,x\,|\,\sqrt{-q}}\right], \quad q = 4\,cd - b^2, \quad (d < 0). \quad (1.15)$$

Integrals (1.9), (1.10), and (1.11) are specific cases of a more general relationship. If you have taken a course in mathematical physics or quantum physics, you might be familiar with a family of polynomials known as *Legendre polynomials.* These are usually designated with the notation $P_n(x)$, indicating a polynomial of order $n$ and argument $x$; the order is restricted to $n \geqslant 0$. The general relationship is

$$\int_0^{\pi} \frac{d\phi}{(1 - \varepsilon \cos \phi)^{n+1}} = \frac{\pi}{(1 - \varepsilon^2)^{(n+1)/2}} P_n(x), \quad (1.16)$$

where the argument $x$ is

$$x = \frac{1}{\sqrt{1 - \varepsilon^2}}. \quad (1.17)$$

Table 1.1 lists the first few Legendre polynomials; you might want to check the first three integrals.

Notice that all of the powers of $x$ in $P_n(x)$ are even (odd) if $n$ is even (odd). If you need higher-order expressions, these polynomials satisfy a recursion relation:

$$(n + 1)P_{n+1}(x) = (2n + 1)\, x\, P_n(x) - nP_{n-1}(x). \quad (1.18)$$

Another occasionally useful integral along the lines of (1.16) is

$$\int_0^{\pi} (1 - \varepsilon \cos \phi)^n\, d\phi = \pi\, (1 - \varepsilon^2)^{n/2}\, P_n(x), \quad (1.19)$$

with the same argument $x$.

**Table 1.1.** Legendre Polynomials.

| $n$ | $P_n(x)$ |
|---|---|
| 0 | 1 |
| 1 | $x$ |
| 2 | $\frac{1}{2}(3x^2 - 1)$ |
| 3 | $\frac{1}{2}(5x^3 - 3x)$ |
| 4 | $\frac{1}{8}(35x^4 - 30x^2 + 3)$ |
| 5 | $\frac{1}{8}(63x^5 - 70x^3 + 15x)$ |

The following binomial expansions will be useful. These all apply for $x^2 < 1$. Expression (1.23) is a general version of the first three; $n$ is unrestricted.

$$\frac{1}{(1 \pm x)} = 1 \mp x + x^2 \mp x^3 + x^4 + \cdots. \tag{1.20}$$

$$\frac{1}{(1 \pm x)^2} = 1 \mp 2x + 3x^2 \mp 4x^3 + 5x^4 + \cdots. \tag{1.21}$$

$$\sqrt{1 \pm x} = 1 \pm \frac{1}{2}x - \frac{1}{8}x^2 \pm \frac{1}{16}x^3 - \frac{5}{128}x^4 + \cdots. \tag{1.22}$$

$$(1 \pm x)^n = 1 \pm nx + \frac{n(n-1)}{2!}x^2 \pm \frac{n(n-1)(n-2)}{3!}x^3 + \cdots. \tag{1.23}$$

# Chapter 2

# Dynamical Quantities in Polar Coordinates

## 2.1 Position, Velocity, Acceleration, Angular Momentum, Torque, and Energy

From Figure 1.1, the position vector $\boldsymbol{r}$ of an object located at polar coordinates $(r, \phi)$ can be written as

$$\boldsymbol{r} = r\hat{\boldsymbol{r}}, \tag{2.1}$$

where the unit vector $\hat{\boldsymbol{r}}$ is given by the first of Equations (1.3).

The velocity of the object is given by the rate of change of its position:

$$v = \frac{d\boldsymbol{r}}{dt} = \dot{r}\hat{\boldsymbol{r}} + r\frac{d\hat{\boldsymbol{r}}}{dt}, \tag{2.2}$$

where we have used the dot notation for $(dr/dt)$; $\dot{r}$ is known as the *radial velocity*. This expression brings out the reason for setting up the time-derivatives of the polar-coordinate unit vectors in the previous chapter. Using that for $(d\hat{\boldsymbol{r}}/dt)$ brings Equation (2.2) to

$$v = \dot{r}\hat{\boldsymbol{r}} + (r\dot{\phi})\,\hat{\boldsymbol{\phi}}. \tag{2.3}$$

The $\hat{\boldsymbol{\phi}}$ component in this expression is the azimuthal *tangential* velocity.

Similarly, but more involved, acceleration is the time-derivative of velocity. Differentiating (2.3) gives

$$\boldsymbol{a} = \ddot{r}\hat{\boldsymbol{r}} + \dot{r}\left(\frac{d\hat{\boldsymbol{r}}}{dt}\right) + (\dot{r}\dot{\phi})\,\hat{\boldsymbol{\phi}} + (r\ddot{\phi})\,\hat{\boldsymbol{\phi}} + (r\dot{\phi})\,\frac{d\hat{\boldsymbol{\phi}}}{dt}. \tag{2.4}$$

Invoking Equations (1.7) and (1.8) reduces this, after more algebra, to

$$\boldsymbol{a} = \{\ddot{r} - r(\dot{\phi})^2\}\hat{\boldsymbol{r}} + \{2\dot{r}\dot{\phi} + r\ddot{\phi}\}\hat{\boldsymbol{\phi}}. \tag{2.5}$$

doi:10.1088/2514-3433/acb430ch2

If the object being tracked is of mass $m$, then the force $\boldsymbol{F}$ on it must be $\boldsymbol{F} = m\boldsymbol{a}$.

A very important quantity will be the angular momentum $\boldsymbol{L}$ of the object. From fundamental physics this is given by

$$\boldsymbol{L} = \boldsymbol{r} \times \boldsymbol{p}, \tag{2.6}$$

where $\boldsymbol{p} = m\boldsymbol{v}$ is the linear momentum $m\boldsymbol{v}$ of the object. From Equations (2.1) and (2.3),

$$\boldsymbol{L} = r\hat{\boldsymbol{r}} \times m[\dot{r}\hat{\boldsymbol{r}} + (r\dot{\phi})\,\hat{\boldsymbol{\phi}}] = (mr^2\,\dot{\phi})(\hat{\boldsymbol{r}} \times \hat{\boldsymbol{\phi}}) = (mr^2\,\dot{\phi})\hat{\boldsymbol{z}}. \tag{2.7}$$

The first term in the cross-product vanished because $\hat{\boldsymbol{r}} \times \hat{\boldsymbol{r}}$ is zero. This expression for $\boldsymbol{L}$ will prove to be *extremely* useful.

Another physics quantity is the torque $\boldsymbol{\tau}$ exerted on the object due to the presence of some external force $\boldsymbol{F}$. An expression for this can be developed in two equivalent ways: by invoking the fundamental definition $\boldsymbol{\tau} = \boldsymbol{r} \times \boldsymbol{F} = \boldsymbol{r} \times m\boldsymbol{a}$, or by recalling that $\boldsymbol{\tau} = d\boldsymbol{L}/dt$. The first of these is a little easier, but you should show that taking $d\boldsymbol{L}/dt$ from Equation (2.7) gives the same result, namely

$$\boldsymbol{\tau} = m(2r\dot{r}\dot{\phi} + r^2\ddot{\phi})\hat{\boldsymbol{z}}. \tag{2.8}$$

We will not use this equation directly; it is recorded here purely for sake of completeness.

The final dynamical quantity we will need is the energy of the moving object, which will be the sum of its kinetic energy $mv^2/2$ and its potential energy. The potential energy will depend on what force it is subject to, so for the moment potential energy will just be represented by the symbol $V$. The kinetic energy can be obtained from Equation (2.3) using $v^2 = \boldsymbol{v} \cdot \boldsymbol{v}$:

$$\frac{1}{2}mv^2 = \frac{1}{2}m(\boldsymbol{v} \cdot \boldsymbol{v}) = \frac{1}{2}m[(\dot{r})^2 + (r\dot{\phi})^2].$$

The symbol $E$ is used for total energy:

$$E = \frac{1}{2}m[(\dot{r})^2 + (r\dot{\phi})^2] + V. \tag{2.9}$$

Readers who are interested in extending these concepts to three dimensions by using spherical coordinates should consult the appendix.

## 2.2 Uniform Circular Motion: A Specific Case of the Acceleration Formula

Equation (2.5) looks rather intimidating at first glance, but contains within itself a familiar result: the centripetal acceleration accompanying the circular motion of an object moving with constant speed in an orbit of radius $r$ in the $xy$-plane. If the motion has a constant radius, then $\dot{r} = \ddot{r} = 0$. The condition of uniform speed demands $\dot{\phi}$ = constant ($\dot{\phi}$ is usually written as $\omega$ in elementary physics texts), and $\ddot{\phi} = 0$. Putting these conditions into Equation (2.5) gives

$$\boldsymbol{a} = -r(\dot{\phi})^2\hat{\boldsymbol{r}}. \tag{2.10}$$

The velocity, Equation (2.3), similarly reduces to

$$v = (r\dot{\phi})\,\hat{\boldsymbol{\phi}}. \tag{2.11}$$

This means that the magnitude of the velocity must be $v = (r\,\dot{\phi})$, or $\dot{\phi} = v/r$. This result, when substituted into Equation (2.10), gives the centripetal acceleration in the familiar from

$$\boldsymbol{a} = -\left(\frac{v^2}{r}\right)\hat{\boldsymbol{r}}. \tag{2.12}$$

It is worth emphasizing that while this gives the acceleration corresponding to uniform circular motion, the nature of the force which gives rise to this acceleration is entirely another question. It might happen that the force is of the familiar gravitational form $\boldsymbol{F} = -(GMm/r^2)\hat{\boldsymbol{r}}$ as we will use in the following chapter, but it might have some other form, say $\boldsymbol{F} = -(k\ r^5)\hat{\boldsymbol{r}}$, where $k$ is some constant that makes the units work out correctly. This would correspond to an attractive force which varies as the fifth power of the distance from the force center. No such force is known to physics, but there is no reason a priori that gravitation must be an inverse-square force. This issue will be explored further in Sections 3.7 and 5.8.

# Chapter 3

# Central Forces

This lengthy chapter explores the physics of central forces. Gravity is the quintessential central force: it acts directly along the line joining the centers of mass of two objects. I know that you know that its magnitudes acts like $F = GMm/r^2$, and we will eventually use this prescription. But to begin with, I will establish a number of very general relations that are valid for any central force acting between two masses $M$ and $m$ whose magnitude behaves as $F(r)$, and get to the specific case of gravity later on. Central means that the force has no "tangential" component, that is, no $\hat{\boldsymbol{\phi}}$ component, only a radial $\hat{\boldsymbol{r}}$ component, and further that the radial component can depend *only* on $r$, that is, $\boldsymbol{F} = F(r)\hat{\boldsymbol{r}}$. Terminology: some authors use the term "central force" for one which has only a radial component but where the radial component might depend on *both* $r$ and $\phi$, that is, a force of the form $\boldsymbol{F} = F(r, \phi)\hat{\boldsymbol{r}}$. If the radial component is further restricted to beginning of the form $\boldsymbol{F} = F(r)\hat{\boldsymbol{r}}$, it is then called a "spherically symmetric" force. There is no universal convention on this; in this book, when I use the term "central force," I strictly mean one of the $\boldsymbol{F} = F(r)\hat{\boldsymbol{r}}$ type. Gravity is of this form.

## 3.1 The Center of Mass and the Reduced Mass

Figure 3.1 shows two masses, $M$ and $m$, lying in the plane of the page (the $xy$-plane). $M$ is intended to be the larger mass, but this need not be the case; the mathematical symbols have no idea what numbers will eventually be assigned to them. Relative to the origin, the location of $M$ is $\boldsymbol{d}_M$ and the location of $m$ is $\boldsymbol{d}_m$. The origin can be any convenient "fixed" reference point; in astronomical terms it might be the origin of the right ascension and declination system, or just the location of a star that is effectively stationary compared to the two-mass system being observed.

I will first remind you of the concept of center of mass (CM) from basic mechanics; this will be a stepping-stone to a more convenient coordinate arrangement for the two-mass system.

doi:10.1088/2514-3433/acb430ch3

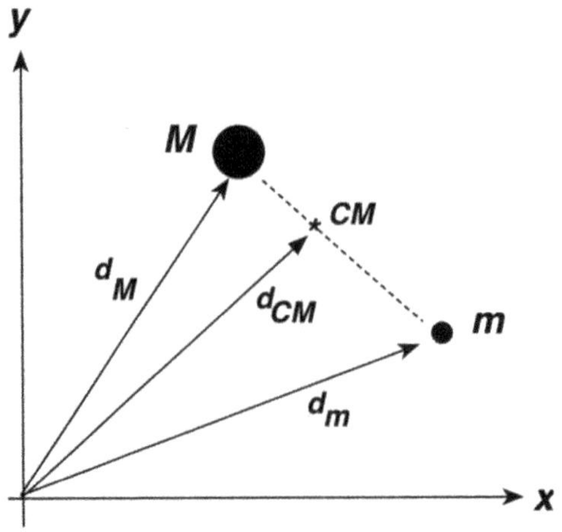


**Figure 3.1.** Two masses, $M$ and $m$, in the $xy$-plane. CM marks the position of their center of mass. Vectors $\boldsymbol{d}_M$, $\boldsymbol{d}_m$, and $\boldsymbol{d}_{\text{CM}}$, respectively, go from the origin to $M$, $m$, and the center of mass.

The position $\boldsymbol{d}_{\text{CM}}$ of the CM is defined as

$$(M + m)\boldsymbol{d}_{\text{CM}} = M\boldsymbol{d}_M + m\boldsymbol{d}_m. \tag{3.1}$$

Taking the second time-derivative of this expression gives the acceleration of the CM in terms of the accelerations of $M$ and $m$:

$$(M + m)\boldsymbol{a}_{\text{CM}} = M\boldsymbol{a}_M + m\boldsymbol{a}_m. \tag{3.2}$$

However, the mass of an object times its acceleration is just the total force acting on it:

$$(M + m)\boldsymbol{a}_{\text{CM}} = \boldsymbol{F}_M + \boldsymbol{F}_m. \tag{3.3}$$

Here is the important point: if the two masses are otherwise isolated from any external force, the only forces acting on them can be those due to each other. But by Newton's third law these must be equal and opposite, so $\boldsymbol{F}_M + \boldsymbol{F}_m = 0$, that is, *the CM can have no acceleration*. This means that the velocity of the CM of the system must be constant; at most, it can be moving with a constant velocity relative to the "external" origin of the $xy$ system.

The value of this is that we can imagine an observer riding along with the CM of the system, effectively giving the problem a new origin point. This sets us up for the next part of the analysis.

Figure 3.2 shows a detailed view of the two masses. Gravitationally, $M$ and $m$ attract each other with equal and opposite forces along the line joining them. Three vectors are of interest here. The vector going from $M$ to $m$ is $\boldsymbol{r}$; the magnitude of this, $r$, is the separation between the two masses at the moment shown. The vectors $\boldsymbol{r}_M$ and $\boldsymbol{r}_m$ go from the center of mass to $M$ and $m$, respectively. In the analysis which follows, $\hat{\boldsymbol{r}}$ is a unit vector pointing from $M$ to $m$, that is, $\hat{\boldsymbol{r}} = \boldsymbol{r}/r$, as usual.

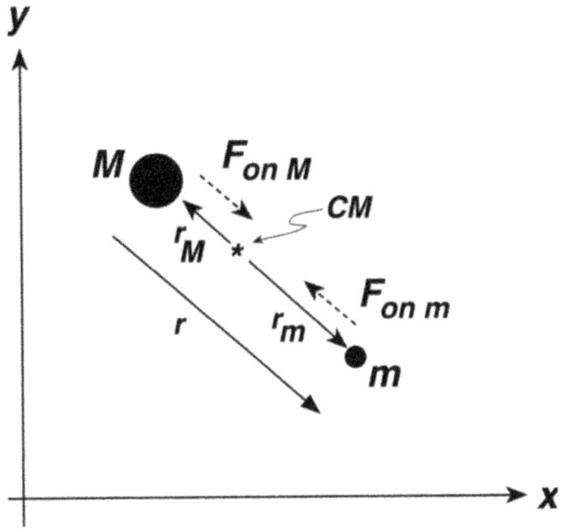


**Figure 3.2.** Detailed view of Figure 3.1. $M$ and $m$ attract each other with equal and opposite forces along the line joining them: Newton's third law. The vector going from $M$ to $m$ is $\boldsymbol{r}$. $M$ is intended to be the larger mass, but this need not be the case. The vectors $\boldsymbol{r}_M$ and $\boldsymbol{r}_m$ go from the CM to $M$ and $m$, respectively. In the analysis in the text, $\hat{\boldsymbol{r}}$ is a unit vector pointing from $M$ to $m$, that is, $\hat{\boldsymbol{r}} = \boldsymbol{r}/r$.

We now make yet another change-of-origin. This may seem overly complicated, but will prove to be extremely convenient. Since we will be concerned with the motions of these masses around each other, it is helpful to regard one of them as being at the origin; in this way the analysis can be divorced from having to specify the CM or some external origin. I choose $M$ to be the new origin; one could equally well choose $m$. From the vectors described above, the force exerted on $M$ by $m$ will be in the direction $+\hat{\boldsymbol{r}}$, and the force exerted on $m$ by $M$ will be in the direction $-\hat{\boldsymbol{r}}$; convince yourself of the veracity of these signs.

If the *magnitude* of the equal-and-opposite force between the two masses is written as $F(r)$, then we can use Newton's second law to write their accelerations with respect to the CM as

$$M\frac{d^2\boldsymbol{r}_M}{dt^2} = F(r)\hat{\boldsymbol{r}} \tag{3.4}$$

and

$$m\frac{d^2\boldsymbol{r}_m}{dt^2} = -F(r)\hat{\boldsymbol{r}}. \tag{3.5}$$

You should check that the signs here are consistent with a mutually-attractive force. Eventually we will put $F(r) = +GMm/r^2$ (remember that $F(r)$ is the magnitude of the force; the unit vectors take care of the directions), but for the present purposes this need not be specified. Now, from the vectors in the figure, we can put things in terms of an $M$-centric system by observing that

$$\boldsymbol{r} = \boldsymbol{r}_m - \boldsymbol{r}_M. \tag{3.6}$$

Taking the second derivative of this expression with respect to time gives the acceleration of $m$ as witnessed by $M$:

$$\frac{d^2\mathbf{r}}{dt^2} = \frac{d^2\mathbf{r}_m}{dt^2} - \frac{d^2\mathbf{r}_M}{dt^2}. \tag{3.7}$$

Now solve Equations (3.4) and (3.5) for the second derivatives of the position vectors and substitute them into Equation (3.7). This gives

$$\frac{d^2\mathbf{r}}{dt^2} = -F(r)\left[\frac{1}{M} + \frac{1}{m}\right]\hat{\mathbf{r}} = -F(r)\left[\frac{M+m}{Mm}\right]\hat{\mathbf{r}}. \tag{3.8}$$

However, $-F(r)\,\hat{\mathbf{r}}$ is the force on $m$:

$$\text{Force on } m = -F(r)\,\hat{\mathbf{r}} = \left(\frac{M\,m}{M+m}\right)\frac{d^2\mathbf{r}}{dt^2} = \mu\mathbf{a}. \tag{3.9}$$

In words, Equation (3.9) says that the force on mass $m$ is equivalent to that acting on a "composite" mass $Mm/(M+m)$, if we take mass $M$ to be the origin of the two-mass system. This is equivalent to saying that we can now regard the system as comprising a stationary mass $M$ located at the origin which is being orbited by another mass, $Mm/(M+m)$, such that the force exerted on this latter mass depends only on the distance $r$ between the two. You may need to take a few minutes to think about these statements before proceeding.

The composite mass $Mm/(M+m)$ is known as the "reduced mass" of the two-body system, and is always designated with the letter $\mu$:

$$\mu = \left(\frac{M\,m}{M+m}\right). \tag{3.10}$$

The "reduced" terminology alludes to two features: the system has been reduced to one mass in motion instead of two, and, if $M \gg m$ as is usually the case for the Sun and a planet, then the reduced mass will be slightly reduced from $m$.

The concept of a reduced mass is so powerful that its significance bears reiterating. By using the reduced mass, we can treat the mutual motions within an isolated two-body system by considering only the motion of one body about another, with one of them imagined to be at a fixed location. This means that in the various dynamical equations of the previous section where the "moving" mass $m$ appears, $m$ must be replaced with $\mu$, but this is a small price to pay for the conceptual simplification gained. Hereafter, we will generally regard $M$ as being at the origin of a polar-coordinate system.

If you do wish to know the positions of the masses relative to the original $xy$ coordinate system, this can be done as follows. Solving $\mathbf{d}_M + \mathbf{r} = \mathbf{d}_m$ for $\mathbf{d}_M$ and substituting into Equation (3.1) gives

$$(M+m)\mathbf{d}_{\text{CM}} = M(\mathbf{d}_m - \mathbf{r}) + m\mathbf{d}_m = (M+m)\mathbf{d}_m - M\mathbf{r},$$

or

$$\boldsymbol{d}_m = \boldsymbol{d}_{\text{CM}} + \left(\frac{M}{M+m}\right)\boldsymbol{r}. \tag{3.11}$$

Conversely, solving Equation (3.6) for $\boldsymbol{r}_m = \boldsymbol{r}_M + \boldsymbol{r}$ and substituting into Equation (3.1) gives

$$\boldsymbol{d}_M = \boldsymbol{d}_{\text{CM}} - \left(\frac{m}{M+m}\right)\boldsymbol{r}. \tag{3.12}$$

This latter expression shows that if $M \gg m$, then mass $M$ will be at essentially the CM of the system; for the solar system, the CM of the system lies below the surface of the Sun (see Section 8.8). But from here on out, our perspective is that $M$ is at the origin, even though to an external observer both $M$ and $m$ will be seen to orbit the CM.

## 3.2 Central Force Dynamics: The Potential

Equation (2.9) in Section 2.1 expresses the total energy of an object in the language of polar coordinates. This includes a term for the potential energy, $V$. This section elaborates on how $V$ can be explicitly formulated.

To begin, recall the work-energy theorem of fundamental mechanics: that the work done on an object by a force $\boldsymbol{F}$ can be expressed as a path integral of the force over the trajectory followed by the object from some initial location $i$ to some final location $f$, and is equal to the change in kinetic energy $K$ of the object:

$$W = \int_i^f \boldsymbol{F} \cdot d\boldsymbol{r} = K_f - K_i. \tag{3.13}$$

For sake of completeness, a proof of this is given here. Since $\boldsymbol{F} = m\boldsymbol{a}$ and $\boldsymbol{a} = d\boldsymbol{v}/dt$, we can write the work integral as

$$W = m \int_i^f \left(\frac{d\boldsymbol{v}}{dt}\right) \cdot d\boldsymbol{r}.$$

The $dt$ in this expression is a scalar, and there is no harm in moving it under $d\boldsymbol{r}$:

$$W = m \int_i^f d\boldsymbol{v} \cdot \left(\frac{d\boldsymbol{r}}{dt}\right)$$

Now, $d\boldsymbol{r}/dt = \boldsymbol{v}$, so this becomes

$$W = m \int_i^f (d\boldsymbol{v} \cdot \boldsymbol{v}).$$

The integrand here can be written as the differential of the square of the magnitude of the velocity:

$$d(v^2) = d(\boldsymbol{v} \cdot \boldsymbol{v}) = (d\boldsymbol{v} \cdot \boldsymbol{v} + \boldsymbol{v} \cdot d\boldsymbol{v}) = 2(d\boldsymbol{v} \cdot \boldsymbol{v}),$$

where the last step follows because dot products are commutative. This renders the work integral as

$$W = \frac{m}{2}\int_i^f d(v^2) = \frac{m}{2}\left(v_f^2 - v_i^2\right),$$

precisely the work-energy theorem.

Return to Equation (3.13). If we are dealing with a system where mechanical energy is conserved, then the change in kinetic energy $K_f - K_i = \Delta K$ must be equal to the negative of the change in potential energy, $-\Delta V$. Hence we can equivalently write the work-energy theorem as

$$W = \int_i^f \boldsymbol{F} \cdot d\boldsymbol{r} = -\Delta V. \tag{3.14}$$

If you have taken a class in advanced calculus, you will know that the change in a function can be expressed as a path integral of its gradient:

$$\Delta V = \int_i^f (\nabla V) \cdot d\boldsymbol{r}. \tag{3.15}$$

Compare Equations (3.14) and (3.15): they are both expressions for $\Delta V$, albeit with a negative sign appearing in (3.14). We must conclude that the force is the negative gradient of the potential energy function:

$$\boldsymbol{F} = -\nabla V. \tag{3.16}$$

This result gives us a recipe for figuring out the potential energy function corresponding to any force. In the case of polar coordinates, the full three-dimensional gradient is

$$\nabla V = \left(\frac{\partial V}{\partial r}\right)\hat{\boldsymbol{r}} + \left(\frac{1}{r}\frac{\partial V}{\partial \phi}\right)\hat{\boldsymbol{\phi}} + \left(\frac{\partial V}{\partial z}\right)\hat{\boldsymbol{z}}. \tag{3.17}$$

For a central force that depends only on the radial distance (gravity!), $\boldsymbol{F} = F(r)\hat{\boldsymbol{r}}$. Consequently, $V$ can have no dependence on $\phi$ or $z$, which means that $F(r) = -(\partial V/\partial r)$. Hence we have

$$V(r) = -\int F(r)dr. \tag{3.18}$$

There will be a constant of integration to be determined, but we will worry about this later.

There is a loose end to be tied up here: how do we know that $\boldsymbol{F}$ is a conservative force, that is, that mechanical energy will be conserved and that we are safe in asserting that $\boldsymbol{F} = -\nabla V$? This draws on another theorem from advanced calculus, namely that if the curl of a vector function is equal to zero ($\nabla \times \boldsymbol{F} = 0$), then you are guaranteed to be able to write that function as the gradient of a scalar function.

This is because the curl of the gradient of any scalar function is always zero: $\nabla \times \nabla V = 0$. So, if $\nabla \times \boldsymbol{F} = 0$, you can write $\boldsymbol{F} = -\nabla V$. In polar coordinates, the curl appears as

$$\nabla \times \boldsymbol{F} = \left[\frac{1}{r}\frac{\partial F_z}{\partial \phi} - \frac{\partial F_\phi}{\partial z}\right]\hat{\boldsymbol{r}} + \left[\frac{\partial F_r}{\partial z} - \frac{\partial F_z}{\partial r}\right]\hat{\boldsymbol{\phi}} + \frac{1}{r}\left[\frac{\partial}{\partial r}(rF_\phi) - \frac{\partial F_r}{\partial \phi}\right]\hat{\boldsymbol{z}}, \tag{3.19}$$

where $(F_r, F_\phi, F_z)$ are the components of $\boldsymbol{F}$ in the form

$$\boldsymbol{F} = F_r\hat{\boldsymbol{r}} + F_\phi\hat{\boldsymbol{\phi}} + F_z\hat{\boldsymbol{z}}. \tag{3.20}$$

Note that some or all of $(F_r, F_\phi, F_z)$ can be functions of some or all of $(r, \phi, z)$. A central force is one where only $F_r$ is non-zero, and depends only on $r$. In this case you should be able to convince yourself that every term in Equation (3.19) vanishes, so $\nabla \times \boldsymbol{F} = 0$: gravity (or any central force) *is* a conservative force and thus possesses a corresponding potential energy function.

What is the potential energy function for the gravitational system of two masses, $M$ and $m$, separated by distance $r$ as in Figure 3.1? If we treat $M$ as being located at the coordinate origin, then the force on $m$ is

$$\boldsymbol{F}_{\text{on } m} = -\frac{GMm}{r^2}\hat{\boldsymbol{r}} = -\nabla V = -\left(\frac{\partial V}{\partial r}\right)\hat{\boldsymbol{r}}, \tag{3.21}$$

where the last part of this expression comes from Equation (3.17). Hence

$$V(r) = +GMm\int\frac{dr}{r^2} = -\frac{GMm}{r} + C, \tag{3.22}$$

where $C$ is a constant of integration.

The presence of the constant of integration is a reminder that potential energy is always arbitrary to within some additive constant: It is *differences* in potential energy that are physically relevant as they dictate differences in kinetic energy, which is a measure of work done. An example of this occurs in basic physics when you write the potential energy of a mass $m$ at height $h$ as $mgh$: you have complete liberty to assign the "zero level" of height. In the present case, we need to decide on some value of $r$ as a calibrating value for $V$, and then assign $C$ so that $V$ is equal to some arbitrarily-assigned number of Joules at the calibrating point.

Obviously, there is an infinitude of possible calibrating points and values of $V$. So that the gravitational potential can be written in one uniform way for every system that might be encountered, it is universally conventional to set $V = 0$ at $r = \infty$; this has the consequences that $C = 0$ and that all values of $V$ will be negative. It is further sometimes customary to write the product $-GMm$ in Equation (3.22) as its own symbol, $\kappa$. So, our recipe for gravitational potential energy appears as

$$V(r) = -\frac{GMm}{r} = +\frac{\kappa}{r}. \tag{3.23}$$

Be careful with the negative sign in the definition of $\kappa$: generations of students have lost points on homework assignments and tests by forgetting it. One has to be very careful with signs in gravitational analyses.

For the remainder of this chapter, I will write the potential energy function just as $V(r)$, coming back to the specific formulation of Equation (3.23) later on.

Warning: Equation (3.23) gives the gravitational potential energy of our two-mass system; I call this the "potential energy function." In some areas of your physics studies, notably electromagnetism and quantum mechanics, you will encounter the "potential function," usually stated more simply as just "the potential." In the gravitational case, the "attracting mass" $M$ is thought of as setting up a "field" given by the "potential function" $U(r) = -GM/r$. If the "test mass" $m$ finds itself in the vicinity of $M$, then the system acquires a potential energy $V(r) = mU(r) = -GMm/r$, per (3.23). The field created by $M$ is presumed to exist even if there is no test mass present to experience it; $m$ is known as a "coupling constant," as it is what links the test mass to the external field. From the point of view of $M$, however, it is mass $m$ that sets up a potential field $U(r) = -Gm/r$, so the potential energy becomes $V(r) = MU(r) = -GMm/r$. In the end, it is irrelevant which mass we consider to be the source of the field and which we consider to experience the field. Many derivations involve just the "potential function," but in this book it is the energetics of an orbiter that are important, so I will always stick to the potential energy formulation—but be sure what function a given source is referring to. Unfortunately, there is no universal convention as to whether $V$ or $U$ refers to the potential or the potential energy.

To close this section, I need to address a question that may have occurred to you: why did I not use the reduced mass $\mu$ in writing Equation (3.21)? It is because potential energy is a function only of the masses and their separation; it is not a "dynamical" quantity having anything to do with motions. Physicists think of potential energy as being a property of the geometric arrangement of a system of particles, not of one particle in isolation.

## 3.3 Why an Inverse-Square Law? The Sesquialterate Proportion

Newton's law of gravitation, Equation (3.21), appears in most physics texts as essentially received wisdom handed down from on high. But why might Newton have looked to an inverse-square law? At an intuitive level, an inverse-square law certainly has considerable appeal. If gravitational force somehow emanates from the Sun to act on planets, it is easy to imagine it spreading out with increasing distance. At distance $r$ from the Sun, the force would be spread over a sphere of area $4\pi r^2$, so having $F \propto 1/r^2$ makes some sense. But aside from this and the somewhat disingenuous response that "An inverse-square force predicts an elliptical orbit," Newton himself offered a quantitative rational for why an inverse-square law might be involved.

Newton's analysis was based on a uniform circular motion argument, specifically, a comparison of the centripetal acceleration of the Moon in its assumed circular orbit around the Earth to the familiar gravitational acceleration $g = 9.81 \text{ m s}^{-2}$ at the surface of the Earth.

Suppose that the radius of the Earth is $R_E$ and that the distance to the Moon from the center of the Earth is $d_{\text{moon}}$ as shown in Figure 3.3. If the speed of the Moon in its orbit is $v_{\text{moon}}$, then its centripetal acceleration must be

$$a_{\text{moon}} = \frac{v_{\text{moon}}^2}{d_{\text{moon}}}. \tag{3.24}$$

We can put $v_{\text{moon}}$ in terms of the Moon's orbital period $T_{\text{moon}}$:

$$v_{\text{moon}} = \frac{2\pi d_{\text{moon}}}{T_{\text{moon}}}, \tag{3.25}$$

and hence write $a_{\text{moon}}$ as

$$a_{\text{moon}} = \frac{4\pi^2 d_{\text{moon}}}{T_{\text{moon}}^2}. \tag{3.26}$$

The ratio of $a_{\text{moon}}$ to $g$ is then

$$\frac{a_{\text{moon}}}{g} = \frac{4\pi^2\, d_{\text{moon}}}{gT_{\text{moon}}^2}. \tag{3.27}$$

The average Earth–Moon distance $d_{\text{moon}}$ is about 384,400 km = $3.844 \times 10^8$ m, and the orbital period is about 27.322 days = $2.361 \times 10^6$ s; both of these numbers were reasonably well-known in Newton's time. Hence

$$\frac{a_{\text{moon}}}{g} = \frac{4\pi^2(3.844 \times 10^8\ \text{m})}{(9.81\ \text{m s}^{-2})(2.361 \times 10^6\ \text{s})^2} = 2.776 \times 10^{-4} \sim \frac{1}{3602}. \tag{3.28}$$

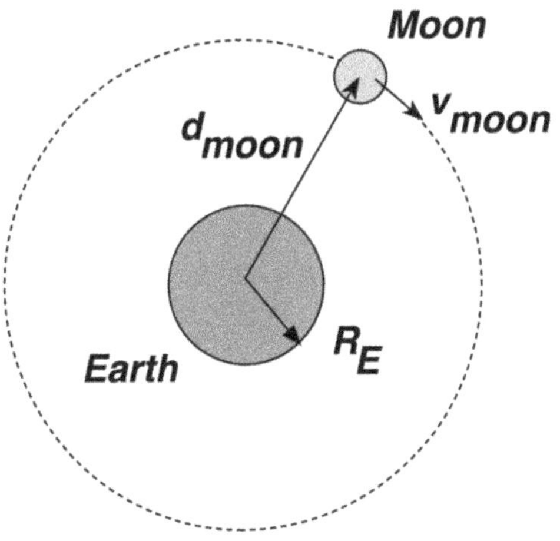


**Figure 3.3.** The Earth and Moon, with the latter in an idealized circular orbit around the former.

Now consider the ratio of the squares of the average radius of the Earth, 6371 km, and the distance to the Moon:

$$\left(\frac{R_E}{d_{moon}}\right)^2 = \left(\frac{6.371 \times 10^6 \text{ m}}{3.844 \times 10^8 \text{ m}}\right)^2 = 2.747 \times 10^{-4} \sim \frac{1}{3640}. \tag{3.29}$$

The striking similarity of these numbers led Newton to surmise that

$$\frac{a_{moon}}{g} = \left(\frac{R_E}{d_{moon}}\right)^2. \tag{3.30}$$

In words: the gravitational acceleration in each case—and hence the gravitational force—is inversely proportional to the distance from the center of the Earth. The numbers do not agree exactly because the Moon's orbit is not a perfect circle, nor is $g$ constant around the surface of the Earth, which is not a perfect sphere.

To think that gravity extended from the surface of the Earth to the Moon was a remarkable speculation on Newton's part. We can do no better than to quote him directly (slightly edited): "And the same year [1665] I began to think of gravity extending to the orb of the Moon, and having found out how to estimate the force with which a globe revolving within a sphere presses the surface of the sphere, from Kepler's rule of the periodical times of the planets being in a sesquialterate proportion of their distances from the centers of their orbs, I deduced that the forces which keep the planets in their orbs must be reciprocally as the squares of their distances from the centers about which they revolve: and thereby compared the force requisite to keep the Moon in her orb with the force of gravity at the surface of the Earth, and found they answer pretty nearly." (See Westfall, p. 143, Westfall 1980.) By "sesquialterate," Newton means in a ratio of three to two: $T \propto d^{3/2}$. This involves Kepler's third law, which is explored in Sections 5.4 and 5.8.

Here is an exercise you should think about: develop an argument as to why the numerator in Equation (3.21) is the product $Mm$ as opposed to the sum $(M + m)$. Hint: think about $\boldsymbol{F} = m\boldsymbol{a}$ as applied to each mass and remember Newton's third law.

## 3.4 Central Force Dynamics: Conservation of Angular Momentum

The first very important consequence that emerges from the fact that any central force can be written in the form $\boldsymbol{F} = F_r\hat{\boldsymbol{r}}$ concerns the angular momentum of our two-body system. Refer again to Figure 3.2. The torque that the "central" mass $M$ exerts on the moving one is given by

$$\boldsymbol{\tau} = \boldsymbol{r} \times \boldsymbol{F} = r\,\hat{\boldsymbol{r}} \times F_r\hat{\boldsymbol{r}} = (rF_r)\,(\hat{\boldsymbol{r}} \times \hat{\boldsymbol{r}}) = 0. \tag{3.31}$$

The zero result follows because any unit vector crossed into itself always gives zero. This result means that a central force can exert no torque, hence the angular momentum of the system must be conserved! From Equation (2.7), this means, on substituting the reduced mass for the orbiting mass, that

$$\boldsymbol{L} = (\mu r^2\, \dot{\phi})\hat{z}. \tag{3.32}$$

Usually, conserved quantities in physics are scalars such as energy or mass. But $\boldsymbol{L}$ is a vector, which means that it has both magnitude and direction. For $\boldsymbol{L}$ to be conserved, both its magnitude and direction must remain constant; once they are set by specifying some initial velocity and position, neither can change. Recall that the magnitude of a vector is always given by the square root of the sum of the squares of its components.

The fact that $\boldsymbol{L}$ is a conserved vector quantity gives us the liberty to specify the plane of the motion of the orbiting mass. The orbiting mass must remain in the chosen plane, otherwise $\boldsymbol{L}$ would change. To see this, look at Figure 3.4, which shows sketches of motions and the directions of their corresponding $\boldsymbol{L}$ vectors, which are given by the right-hand-rule. In each case you could imagine changing the plane of the orbit while keeping $\boldsymbol{L}$ of the same magnitude, but this would violate the conservation of $\boldsymbol{L}$ *as a vector*. This is an important point with which you should feel very comfortable.

The constancy of $\boldsymbol{L}$ means that we can write Equation (3.32) as

$$\boldsymbol{L} = +(\mu r^2\dot{\phi})\hat{z} = \text{constant}. \tag{3.33}$$

*Look* at this expression: it tells us that for motion in the $xy$-plane, $\boldsymbol{L}$ must point along the $z$-axis, consistent with the right-hand-rule. Whether $\boldsymbol{L}$ points toward $+z$ or $-z$ depends on the sign of $\dot{\phi}$: toward $+z$ if counterclockwise, toward $-z$ if clockwise.

The magnitude of $\boldsymbol{L}$ must be

$$L = \mu\, r^2\dot{\phi} = \text{constant}. \tag{3.34}$$

A rearrangement of this results leads to an *extremely valuable expression.* Isolate $\dot{\phi}$ and write it explicitly in differential form:

$$\dot{\phi} = \frac{d\phi}{dt} = \frac{L}{\mu r^2}. \tag{3.35}$$

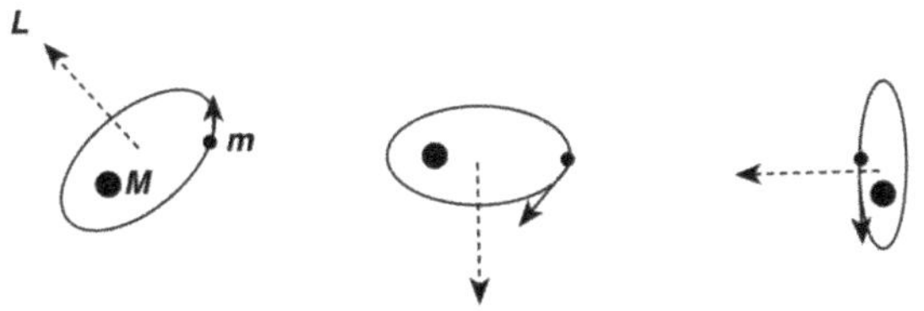


**Figure 3.4.** Sketches of a mass $m$ orbiting a larger mass $M$ in three different orbital planes. The solid arrows show the direction of motion, and the dashed arrows the direction of the angular momentum vectors: use the right-hand-rule. The orbit in the leftmost sketch should be imagined as tilted at some angle with respect to the plane of the page, with $m$ moving into the page. The center sketch is intended to represent an orbital plane exactly perpendicular to the plane of the page, as is the rightmost one: these last two orbital planes are to be imagined as perpendicular to each other. Any alteration in any of the orbital planes, even if the speeds of the motions are kept constant, would change $\boldsymbol{L}$ because it is a vector.

The value of this expression is that it will enable us to change integrals over $d\phi$ to ones over $dt$ and vice-versa. We will use this expression frequently for exactly this purpose:

$$d\phi = \left(\frac{L}{\mu r^2}\right)dt \iff dt = \left(\frac{\mu r^2}{L}\right)d\phi. \tag{3.36}$$

As a first application of this expression, look back to Equation (2.9) for the total energy of the system. Invoke Equation (3.35) to replace $\dot{\phi}$, replace $m$ with the reduced mass $\mu$, and write the potential energy as $V(r)$ as a reminder that we are dealing with a central force; the explicit $\kappa/r$ form of Equation (3.23) will be used later. Also, for a reason that will become apparent in the following section, write the radial velocity in explicit derivative form:

$$E = \frac{1}{2}\left[\mu\left(\frac{dr}{dt}\right)^2 + \frac{L^2}{\mu r^2}\right] + V(r). \tag{3.37}$$

The advantage of writing $E$ in this way is that since both $E$ and $L$ are constants, there are fewer variable quantities to contend with. The values of $E$ and $L$ are ultimately set by initial conditions.

## 3.5 Central Force Dynamics: Integrals of the Motion

In this section, we establish three integrals for central-force motions that link time, the azimuthal position $\phi$, and the radial distance $r$. All of these expressions derive from the expressions for total energy and angular momentum, Equations (3.36) and (3.37).

Begin by rearranging (3.37) to isolate the radial velocity $(dr/dt)$:

$$\left(\frac{dr}{dt}\right) = \pm\sqrt{\frac{2}{\mu}\left[E - V(r) - \frac{L^2}{2\mu r^2}\right]}. \tag{3.38}$$

Separating variables gives:

$$dt = \frac{dr}{\sqrt{(2/\mu)[E - V(r) - L^2/2\mu r^2]}}. \tag{3.39}$$

Strictly, there are two possible solutions here because of the square root; the choice of which sign to use depends on the situation intended. If we imagine that at some time the moving mass is at radial distance $r_0$, we can integrate Equation (3.39) to some later general position $r$ at time $t$; presuming that $t > t_o$ leads to choosing the positive sign:

$$t - t_o = \int_{r_0}^{r} \frac{dr}{\sqrt{(2/\mu)[E - V(r) - L^2/2\mu r^2]}}. \tag{3.40}$$

This is the first integral of the motion. If this integral can be solved for a given potential energy function $V(r)$, we will have an expression for $t(r)$. Whether or not the result can be rearranged to write an expression for $r(t)$ depends on the complexity of $t(r)$; often, this is not possible.

The sum $V(r) + L^2/2\mu r^2$ is often referred to as the effective potential; this will be explored further in Sections 5.9 and 5.10 for the case of the inverse-square force.

The second integral also follows from Equation (3.38), but by transforming $(dr/dt)$ into an angular form and invoking Equation (3.35):

$$\left(\frac{dr}{dt}\right) = \left(\frac{dr}{d\phi}\right)\left(\frac{d\phi}{dt}\right) = \left(\frac{dr}{d\phi}\right)\left(\frac{L}{\mu r^2}\right) = \pm\sqrt{\frac{2}{\mu}\left[E - V(r) - \frac{L^2}{2\mu r^2}\right]}. \tag{3.41}$$

Again separate variables and integrate, with $\phi$ going from $\phi_o$ to a general value $\phi$ while $r$ goes from $r_0$ to $r$. The sense of the change of $\phi$ will be captured in the sign of $L$ (positive for "counterclockwise" motion, negative for "clockwise," although it is the magnitude $L^2$ that appears here), so I again drop the ± sign:

$$\phi - \phi_o = \int_{r_0}^{r} \frac{(L/\mu r^2)}{\sqrt{(2/\mu)[E - V(r) - L^2/2\mu r^2]}}\, dr. \tag{3.42}$$

If the integral can be solved, we will have an expression for $\phi(r)$.

The third integral follows directly from Equation (3.36):

$$\phi - \phi_o = \frac{L}{\mu}\int_{t_o}^{t} \frac{dt}{r^2}. \tag{3.43}$$

The integrand here involves two variables, $r$ and $t$; one would have to know $r(t)$ in order to evaluate it. We will see an ingenuous use of this expression in Section 8.9.

## 3.6 Central Force Dynamics: Acceleration in Terms of the Azimuthal Angle

Here we cast the expression for acceleration in Equation (2.5) into a form that will be very useful in Chapter 5. This expression was

$$\boldsymbol{a} = \{\ddot{r} - r(\dot{\phi})^2\}\hat{\boldsymbol{r}} + \{2\dot{r}\,\dot{\phi} + r\,\ddot{\phi}\,\}\hat{\boldsymbol{\phi}}. \tag{3.44}$$

This could be left as it is for use in a force analysis, but a further simplification can be effected. Kepler's third law tells us planets orbit the Sun in elliptical paths. The shape of an ellipse can be written very compactly in the form $r(\phi)$ (see Equation (4.3)), so it is convenient to have this expression in terms of derivatives with respect to $\phi$ as opposed to with respect to time. This can be accomplished with the conservation of angular momentum Equation (3.35):

$$\dot{\phi} = \frac{L}{\mu\, r^2}. \tag{3.45}$$

Take the time derivative of this, remembering that both $L$ and $\mu$ are constants:

$$\ddot{\phi} = -\frac{2L\dot{r}}{\mu r^3}. \tag{3.46}$$

Be careful to note the factor of $\dot{r}$ here. Now substitute Equations (3.45) and (3.46) into Equation (3.44); you will find that the $\hat{\boldsymbol{\phi}}$ term vanishes completely (a consequence of the conservation of angular momentum), leaving only the radial term:

$$\boldsymbol{a} = \left\{\ddot{r} - \frac{L^2}{\mu^2 r^3}\right\}\hat{\boldsymbol{r}}. \tag{3.47}$$

The $\ddot{r}$ term can be put in terms of $\phi$ as was done in Equation (3.41):

$$\dot{r} = \left(\frac{dr}{dt}\right) = \left(\frac{dr}{d\phi}\right)\left(\frac{d\phi}{dt}\right) = \left(\frac{dr}{d\phi}\right)\left(\frac{L}{\mu r^2}\right). \tag{3.48}$$

Taking another time derivative gives

$$\begin{aligned}\ddot{r} &= \frac{d}{dt}\left[\left(\frac{dr}{d\phi}\right)\left(\frac{L}{\mu r^2}\right)\right] = \left[\frac{d}{dt}\left(\frac{dr}{d\phi}\right)\right]\left(\frac{L}{\mu r^2}\right) + \left(\frac{dr}{d\phi}\right)\left[\frac{d}{dt}\left(\frac{L}{\mu r^2}\right)\right] \\ &= \left[\frac{d}{dt}\left(\frac{dr}{d\phi}\right)\right]\left(\frac{L}{\mu r^2}\right) + \left(\frac{dr}{d\phi}\right)\left[-\frac{2L}{\mu r^3}\left(\frac{dr}{dt}\right)\right].\end{aligned} \tag{3.49}$$

In the last term in this expression, $(dr/dt)$ can be replaced with Equation (3.48). In the first term, differentiation with respect to $t$ can be transformed into differentiation with respect to $\phi$ by using another application of the chain rule: that for any function $X$,

$$\frac{dX}{dt} = \frac{dX}{d\phi}\left(\frac{d\,\phi}{dt}\right). \tag{3.50}$$

In our case, the factor of $(d\phi/dt)$ that this manipulation introduces can be dealt with by again invoking Equation (3.35). Making these changes in Equation (3.49) gives

$$\ddot{r} = \left(\frac{d^2r}{d\phi^2}\right)\left(\frac{L}{\mu r^2}\right)^2 - \left(\frac{dr}{d\phi}\right)^2\left(\frac{2L^2}{\mu^2 r^5}\right). \tag{3.51}$$

Substituting this into Equation (3.47) gives, after extracting some common factors, our final expression for $\boldsymbol{a}$:

$$\boldsymbol{a} = \left(\frac{L^2}{\mu^2 r^2}\right)\left[\frac{1}{r^2}\left(\frac{d^2r}{d\phi^2}\right) - \left(\frac{2}{r^3}\right)\left(\frac{dr}{d\phi}\right)^2 - \frac{1}{r}\right]\hat{\boldsymbol{r}}. \tag{3.52}$$

This expression may look awkward, but it contains no time-derivatives. In preparation for material to be covered in Chapter 5, we now combine this result with Equation (3.9) to get an expression for the magnitude of the radial force $F(r)$ responsible for the acceleration, $\boldsymbol{F} = -\mu\boldsymbol{a}$; be careful to note the negative sign in the left side of (3.9):

$$F(r) = -\left(\frac{L^2}{\mu r^2}\right)\left\{\frac{1}{r^2}\left(\frac{d^2r}{d\phi^2}\right) - \frac{2}{r^3}\left(\frac{dr}{d\phi}\right)^2 - \frac{1}{r}\right\}. \tag{3.53}$$

The value of this expression is that if we know a trajectory $r(\phi)$ which we believe to be caused by a central force, we can determine the radial dependence of the force $F(r)$ by working out the right side; $L$ will then be a (constant) parameter in the force law. If we further believe that we have an independent expression for $F(r)$, such as $GMm/r^2$, we will then be able to write down a prescription for $L$ in terms of $G$, $M$, $m$, and whatever parameters appear in the expression for $r(\phi)$. This is exactly what will be done in Chapter 5.

**Exercise:** A planet is moving in the unusual spiral orbit $r = k\phi$, where $k$ is a constant. What is the corresponding force law $F(r)$ in terms of $L$, $\mu$, and $k$?

Answer: $F(r) = (L^2/\mu)(2k^2/r^5 + 1/r^3)$

**Exercise:** For motion described by $r = R_o \exp(k\phi)$ where $R_o$ and $k$ are constants, what is the corresponding force law $F(r)$ in terms of $L$, $\mu$, and $k$?

Answer: $F(r) = (L^2/\mu r^3)(k^2 + 1)$

Finally, in some textbooks you may encounter an alternate but equivalent expression for $\boldsymbol{a}$. This is explained here, although we will not use it. Define a new radial variable, $u = 1/r$. You should be able to prove that Equation (3.53) transforms to (be careful not to confuse $u$ and $\mu$)

$$F(u) = +\frac{L^2u^2}{\mu}\left\{\frac{d^2u}{d\phi^2} + u\right\}. \tag{3.54}$$

This form of the acceleration vector is known as the Binet equation, after French mathematician Jacques Philippe Marie Binet (1786–1856).

## 3.7 Newton's Shell-Point Equivalency Theorem

This section can be considered optional in that it does not deal with an orbital issue per se, but it does treat a historically important question. This proof invokes the use of the spherical coordinates of Appendix A.

Every physics student learns that if a point mass $m$ finds itself outside a uniformly-dense shell of total mass $M$, then the shell can be regarded as acting as a point mass at a distance $r$ from $m$, where $r$ is the distance from the center of the shell to $m$. This is Newton's shell-point equivalency theorem. Once it has been established for a shell, it can be extended to a solid sphere by building the sphere up out of infinitesimally thin shells. Each shell could in fact be of different density, although they must be isotropic, that is, the density cannot vary from place to place within a shell. Establishing shell-point equivalency allowed Newton to treat the Sun and planets as point masses, so it is a cornerstone underlying his gravitational theory and its application to celestial mechanics. This theorem is now so familiar that it is easy to forget just how beautiful and powerful it is. Physics Nobel laureate Richard Feynman captured its magnificence by describing it in almost supernatural terms: "When we add the effects all together, it seems a miracle that the net force is exactly the same as we would get if we put all the mass in the middle!" (Feynman et al. 1963).

This section offers a proof of this theorem; this is adapted from a publication by this author (Reed 2022). While this development is a bit lengthy, this version contains a bonus not present in most textbook treatments: the analysis is more general than just an inverse-square force, and reveals the interesting conclusion that *both* an inverse-square force and the familiar Hooke's law force of a mass-spring system exhibit shell-point equivalency. In addition, we will see that these are the *only* central power-law forces that possess this equivalency.

To analyze this, it is helpful to invoke a combination of spherical and Cartesian coordinates. The general concept is that gravity is presumed to be a central power-law force such that the force exerted by an element of mass $dM$ of a uniform shell on an external test mass $m$ is $\boldsymbol{F} = +Knm(dM)r^{n-1}\hat{\boldsymbol{r}}$, where $r$ is the element-to-test mass distance. The potential energy is then $V(r) = -Km(dM)r^{n}$; you should check that $\boldsymbol{F} = -(\partial V/\partial r)\hat{\boldsymbol{r}}$. $K$ is a constant that would have to be chosen to make the units work out correctly, and the power $n$ is intended to be an integer, although this need not strictly be the case. Note that if $(Kn) > 0$, the force will be repulsive, and if $(Kn) < 0$ it will be attractive; this is the case of practical interest. The units of $K$ will depend on the choice of the power $n$; when $n = -1$, set $K \rightarrow G$ to recover the Newtonian case.

Figure 3.5 shows the setup. A spherical shell of radius $R$ is placed with its center at the origin. Let the shell have mass per square meter $\sigma$. This is known as an areal mass

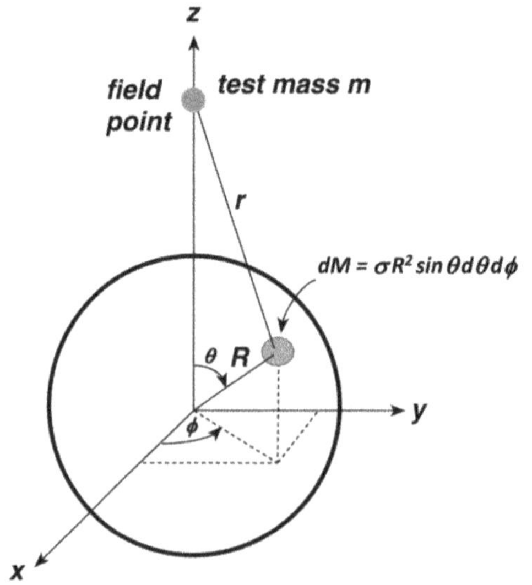


**Figure 3.5.** Isotropic shell of radius $R$ and mass density $\sigma$ per unit area centered on the origin. A field point is located at distance $z$ along the $z$-axis.

density; the Greek letter $\sigma$ is equivalent to “s” and serves as a reminder that we are dealing with a *s*urface area density. For a shell of radius $R$, the total mass $M$ will be $M = 4\pi R^2\sigma$, an expression we will use a little later.

Locate a “field point” at distance $z$ from the center along the $z$-axis; the test mass $m$ is placed there. Because the shell is spherically symmetric, we can put the field point wherever we wish; this choice makes the algebra easier. Also, instead of computing the force on the test mass directly, it is easier to compute the potential energy of the system and then take a negative gradient; this lets us replace a vector integration with a scalar one.

In spherical coordinates, a patch of surface located at $(\theta, \phi)$ has area $dA = R^2 \sin\theta\, d\theta\, d\phi$. The patch will then have mass $dM = \sigma R^2 \sin\theta\, d\theta\, d\phi$. The patch-to-field-point distance is $r$; from the law of cosines this is given by

$$r^2 = z^2 + R^2 - 2zR\cos\theta. \tag{3.55}$$

As described above, the potential energy is written in the generalized power-law form $V = -Km(dM)r^n = -Km\sigma R^2 r^n \sin\theta\, d\theta\, d\phi$. To get the total potential energy, we have to integrate over both $\theta$ and $\phi$ to cover the entire surface:

$$V = -Km\sigma R^2 \int_0^{2\pi}\int_0^{\pi} r^n \sin\theta\, d\theta d\phi. \tag{3.56}$$

The integral over $\phi$ gives $2\pi$ directly:

$$V = -2\pi Km\sigma R^2 \int_0^{\pi} r^n \sin\theta\, d\theta. \tag{3.57}$$

In the expression (3.55) for $r^2$, simplify by setting $a = z^2 + R^2$ and $b = 2zR$, that is, write $r^2 = a - b\cos\theta$. This gives

$$V = -2\pi Km\sigma R^2 \int_0^{\pi} (a - b\cos\theta)^{n/2} \sin\theta\, d\theta. \tag{3.58}$$

Note the power $n/2$; this is because Equation (3.55) gives $r^2$, but the integrand in Equation (3.57) involves $r^n$.

To simplify this integral a change of variable is made, which is to define $w = b\cos\theta$. This gives $dw = -b\sin\theta\, d\theta$, from which we can write $\sin\theta d\theta = -dw/b = -dw/2zR$. This transforms the potential energy to

$$V = \frac{\pi Km\sigma R}{z}\int_b^{-b} (a - w)^{n/2} dw. \tag{3.59}$$

Check that the limits and signs have been transformed correctly.

For $n \neq -2$ (which is discussed below), this integral has the straightforward general form

$$\int (a - w)^{n/2} dw = -\frac{(a - w)^{n/2+1}}{n/2 + 1}. \tag{3.60}$$

Be careful not to lose the negative sign here.

Now replace $a$ with $z^2 + R^2$ and evaluate the limits, setting $b = 2zR$. This gives

$$V = \frac{-2\pi Km\sigma R}{(n+2)z}[(z^2 + R^2 + 2zR)^{(n+2)/2} - (z^2 + R^2 - 2zR)^{(n+2)/2}]. \tag{3.61}$$

We can simplify the terms within the square brackets by writing them as

$$(z^2 + R^2 \pm 2zR)^{1/2} = (z \pm R).$$

This allows us to simplify the potential energy to

$$V = -\frac{KMm}{2(n+2)zR}[(z+R)^{n+2} - (z-R)^{n+2}], \tag{3.62}$$

where $\sigma$ has been replaced by $M/4\pi R^2$ to put things in terms of the total mass of the shell.

Check: For $K = G$ and $n = -1$, $V$ reduces to the standard form $-GMm/z$, where $z$ is playing the role usually written as $r$, the shell center to field point distance.

Assume that $n > -2$. The two round-bracketed terms within the square bracket can be expanded using the binomial theorem,

$$(a + x)^N = \sum_{j=0}^{N} \binom{N}{j} x^j a^{N-j},$$

where the binomial coefficients are given by

$$\binom{N}{j} = \frac{N!}{j!\,(N-j)!},$$

and where the exclamation marks designate factorials. In the present case $N \equiv n + 2$, and this applies as

$$(z \pm R)^{n+2} = \sum_{j=0}^{n+2} \binom{n+2}{j} (\pm R)^j z^{n+2-j}. \tag{3.63}$$

Now, $(\pm R)^j$ can be written as $(\pm 1)^j R^j$. Invoking these compactions renders the potential energy as

$$V = -\frac{KMm}{2(n+2)} \sum_{j=0}^{n+2} \binom{n+2}{j} [1 - (-1)^j]\, z^{n+1-j}\, R^{j-1}, \tag{3.64}$$

where $R^j$ became $R^{j-1}$ via the factor of $R$ in the denominator of the prefactor in (3.62); similarly for $z^{n+2-j}$ becoming $z^{n+1-j}$.

The force on the test mass is then

$$\boldsymbol{F} = -\left(\frac{\partial V}{\partial z}\right)\hat{z} = \frac{KMm}{2(n+2)} \sum_{j=0}^{n+2} \binom{n+2}{j} [1 - (-1)^j](n + 1 - j) z^{n-j}\, R^{j-1}\, \hat{z}. \tag{3.65}$$

The term in square brackets, $1 - (-1)^j$, will be zero if $j$ is even, and will be equal to 2 if $j$ is odd; in this latter case this factor of 2 will cancel the one in the prefactor. Thus, only odd values of $j$ will make any contribution to the sum. Also, from the $(n + 1 - j)$ term, there will be no contribution when $j = n + 1$.

Shell-point equivalency means that all terms in $\boldsymbol{F}$ must be independent of $R$, that is, the force cannot depend on the radius of the shell. The cases of integer values of the power $n$ can be analyzed separately:

(a) If $n = -1$, then the sum involves only $j = 0$ and 1, and the only surviving term will be that for $j = 1$. This gives $\boldsymbol{F} = -(KMm/z^2)\hat{z}$, which is exactly the Newtonian result.

(b) For $n = 0$, then $j = 0$, 1, or 2, but again only $j = 1$ contributes. But here $(n + 1 - j) = 0$, so there is no net force. This is exactly what one would expect for $V$ = constant: the gradient vanishes. This is a mathematically valid but physically rather null case.

(c) For $n = +1$, there are non-vanishing terms in the sum for $j = 1$ and $j = 3$. The $j = 1$ term is independent of $R$ as in case (a) above, but the $j = 3$ term gives a contribution of the form $R^2/z^2$, destroying shell-point equivalency.

(d) For $n = +2$, the mass-spring harmonic potential, the non-vanishing terms in the sum are again $j = 1$ and $j = 3$. In this case, however, the contribution from $j = 3$ vanishes because $(n + 1 - j) = 0$. The $j = 1$ term yields $\boldsymbol{F} = (2KMmz)\hat{z}$, which possesses shell-point equivalency; the factor of 2 arises from the binomial coefficient. If $K > 0$ as with the Newtonian case, this will be a *repulsive* force. This might seem strange, but can be understood via energy conservation. The potential energy in this case, aside from a constant factor involving $R^{-2}$, behaves as $V = -KMmz^2$. A repulsive force will endow a test mass with increasing kinetic energy; as $z$ increases, the potential energy must correspondingly become more negative.

(e) For $n > 2$, there will be terms arising from $j = 1, 3, 5, \ldots$, but terms from the factor of $R^{j-1}$ in Equation (3.65) will survive, destroying shell-point equivalency. As for other possibilities, setting $n = -2$ in Equation (3.59) leads to a logarithmic form for $V$ where terms in $R$ remain. Similarly, for $n < -2$, the terms in the potential cannot be brought to a common denominator without introducing offending $R$-terms. For non-integer values of $n$, the finite binomial expansions will become infinite binomial series, not all of the terms of which can be forced to vanish through the factors of $(n + 1 - j)$, so shell-point equivalency is again impossible. Overall, *only inverse-square and harmonic-oscillator forces possess shell-point equivalency.*

Non inverse-square forces will be revisited briefly in Section 5.8.

## References

Feynman, R. P., Leighton, R. B., & Sands, M. 1963, The Feynman Lectures on Physics (Vol. I,; Reading, MA: Addison-Wesley) (see Section 13)

Reed, B. C. 2022, AmJPh, 90, 394

Westfall, R. S. 1980, Never at Rest: A Biography of Isaac Newton (Cambridge: Cambridge Univ. Press)

# Chapter 4

# The Ellipse

In this brief chapter we review some of the geometrical properties of ellipses in preparation for analyzing orbits in Chapter 5. We also examine how Copernicus and Kepler determined the periods and sizes of planetary orbits, a triumph of classical geometry.

## 4.1 The Ellipse in Polar and Cartesian Coordinates

To a mathematician, an ellipse is the conic section formed when a plane cuts through a cone obliquely to the axis of the cone. Here we will construct equations which describe an ellipse based on a method you may have seem in elementary or high school: looping a string around two tacks, drawing it taut with a pencil, and moving the pencil around while keeping the string taut.

To put this in analytic form, begin by selecting a distance, "$a$" and a second number known as the eccentricity and designated with $\varepsilon$. This latter number is restricted to $0 \leqslant \varepsilon < 1$. Now look at Figure 4.1. Draw a conventional set of $xy$ axes, and place two points on the $x$-axis equidistant from the origin at positions $x = \pm a\varepsilon$. These are known as the foci (singular: focus) of the ellipse. Then pick some starting point $(x, y)$ as shown. The distance from the left focus to this point is designated as $r$, and that from the right focus to the point is labeled $s$. The angle $\phi$ is measured from the $x$-axis counterclockwise to the length $r$, with vertex at the left focus. Strictly, this length is a vector, $\boldsymbol{r}$, the "radial vector."

An ellipse is defined by the locus of points such that the sum of $s$ and $r$ is always equal to $2a$:

$$s + r = 2a. \tag{4.1}$$

Apply the law of cosines to the triangle $s - r - (2a\varepsilon)$:

$$s^2 = r^2 + (2a\varepsilon)^2 - 2(2a\varepsilon)r\cos\phi. \tag{4.2}$$

doi:10.1088/2514-3433/acb430ch4

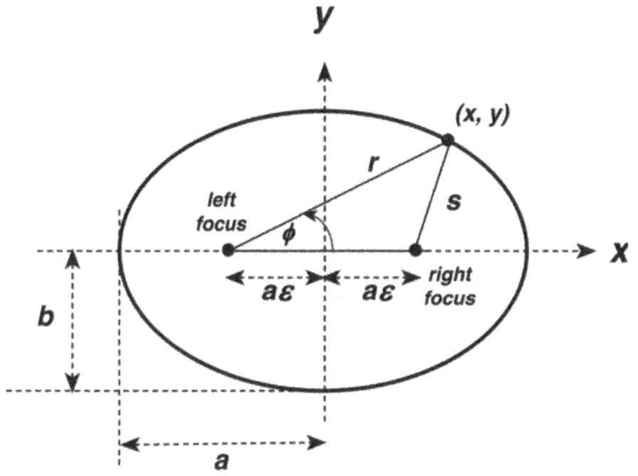


**Figure 4.1.** Construction of an ellipse in the $xy$-plane. The major axis lies along the $x$-axis; the semimajor axis is $a$, and the semiminor axis is $b$, as described in the text. The two foci are located along the semimajor axis, each at distance $a\varepsilon$ from the origin, where $\varepsilon$ is the eccentricity ($0 \leqslant \varepsilon < 1$). The angle $\phi$ is measured from the major axis counterclockwise to the "radial vector" $r$. For most planets the eccentricity is not as great as depicted here, but for comets it can be large.

Now isolate $s$ from Equation (4.1), square it, and substitute into the left side of Equation (4.2):

$$4a^2 + r^2 - 4ar = r^2 + (2a\varepsilon)^2 - 2(2a\varepsilon)r\cos\phi.$$

The factors of $r^2$ cancel; what remains can be solved for $r$ as a function of $\phi$:

$$r = \frac{a(1 - \varepsilon^2)}{1 - \varepsilon\cos\varphi}. \tag{4.3}$$

This is the *polar form* of the equation of an ellipse. Notice that if $\varepsilon \to 0$, the two foci merge at $(x, y) = (0, 0)$, and the ellipse reduces to a circle of radius $a$. A circle is a zero-eccentricity ellipse.

We come now to an important physical point. You might wonder why the left focus was chosen as the reference point for measuring $\phi$, as opposed to the usual convention of having the origin as the reference point. The reason for this is that *Kepler's first law states that all planets orbit the Sun in an elliptical path with the Sun at one focus.* For orbital analysis, it is a focus that is physically important, not the geometric center of the ellipse. *WARNING: Physically, it does not matter if we designate the left or right focus as the "solar" one. I will consistently use the left one, but many authors use the right one. Unfortunately, there is no universal convention for this choice. In texts that use the right focus, all formulae will be identical to those developed in this book except that my* $\varepsilon$ *will appear in right-focus texts as* $-\varepsilon$*, and you have to keep an eye out for this. In a right-focus text, the numerator of Equation (4.3) will be the same as here because* $(-\varepsilon^2) = \varepsilon^2$*, but the denominator will be* $(1 + \varepsilon\cos\phi)$.

The full width of the ellipse along the $x$-axis is $2a$ (check this!). For the ellipse of Figure 4.1, the $x$-axis is known as the major axis, and the $y$-axis is known as the minor axis. $a$ is known as the *semimajor axis*. When $\phi = 0$, $r = a(1 + \varepsilon)$; for an orbit, this is the greatest distance that a planet reaches from the Sun. To astronomers, this point is known as "*aphelion*." Conversely, when $\phi = 180°$, $r = a(1 - \varepsilon)$; this is the point of closest approach, "*perihelion*." For satellites orbiting the Earth, these become "apogee" and "perigee," and for satellites orbiting the Moon one has "apolune" and "perilune." In general, the terms are "apapsis" and "periapsis," and the major axis is called the "line of apsides." In addition to being known as the "azimuthal angle," $\phi$ is also known as the "apsidal angle." Did you get all of these?

The semiminor axis gives the greatest distance that the curve of the ellipse reaches along the $y$-axis (the "minor axis"); this is given by (prove it)

$$b = a\sqrt{1 - \varepsilon^2}. \tag{4.4}$$

The equation of an ellipse can also be written in Cartesian form, with the origin at the conventional $(x, y) = (0, 0)$ location. From Figure 4.1 you should be able to see that the conversion is

$$\begin{aligned} x &= r\cos\varphi - a\varepsilon \\ y &= r\sin\varphi. \end{aligned} \tag{4.5}$$

As an algebra exercise, check that the following holds:

$$\frac{x^2}{a^2} + \frac{y^2}{b^2} = 1. \tag{4.6}$$

This is the Cartesian equation for an ellipse. We will use it in the following section, but not thereafter.

**Exercise:** Perform a binomial expansion on the denominator of Equation (4.3) for $\varepsilon$ small to second order in $\varepsilon$ and use the identity $\cos^2\phi = [1 + \cos(2\phi)]/2$ to show that the equation for an ellipse can be written as $r = C_1 + C_2(\cos\phi) + C_3(\cos 2\phi) + \cdots$. What are $C_1$, $C_2$, and $C_3$?

Answer: $\{C_1,\ C_2,\ C_3\} = \{a(1 - \varepsilon^2)(1 + \varepsilon^2/2),\quad a(1 - \varepsilon^2)\varepsilon,\quad a(1 - \varepsilon^2)(\varepsilon^2/2)\}$.

## 4.2 Area of an Ellipse

In this section we develop a formula for the area of an ellipse. In conjunction with a result established in the following section, this will be useful in an analysis of Kepler's third law in Chapter 5.

Figure 4.2 shows an ellipse with a shaded area of width $dx$ and height $y(x)$ located at distance $x$ from the center.

From the symmetry of the situation, we can express the total area of the ellipse as

$$A_{\text{ellipse}} = 4\int_0^a y(x)dx = 4b\int_0^a \sqrt{1 - x^2/a^2}\,dx = \frac{4b}{a}\int_0^a \sqrt{a^2 - x^2}\,dx. \tag{4.7}$$

The integral itself incorporates only one quadrant of the ellipse; the factor of 4 is to get the total area. This integral is standard, and gives

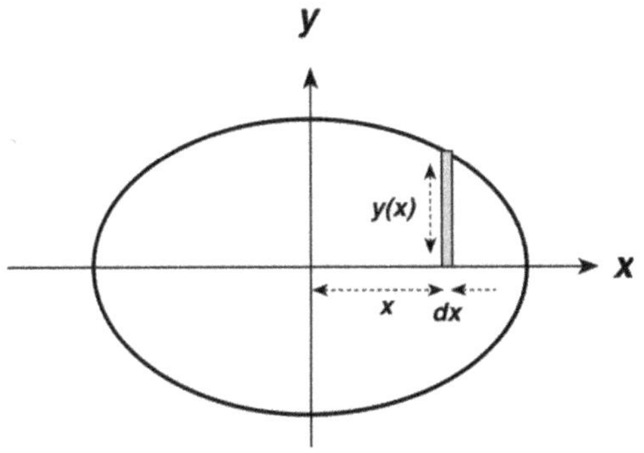


**Figure 4.2.** Construction for the area of an ellipse.

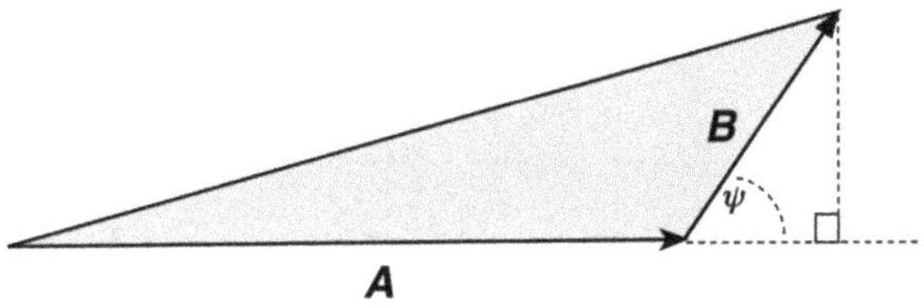


**Figure 4.3.** Construction for the area of a triangle.

$$A_{\text{ellipse}} = \frac{4b}{a}\left[\frac{x}{2}\sqrt{a^2 - x^2} + \frac{a^2}{2}\tan^{-1}\left(\frac{x}{\sqrt{a^2 - x^2}}\right)\right]_0^a. \tag{4.8}$$

Both lower-limit terms evaluate to zero, and the first term is zero at the upper limit. The second term at the upper limit becomes $\tan^{-1}(\infty)$, which is just $\pi/2$. A similar such limit will arise in Chapter 5. The final result is then

$$A_{\text{ellipse}} = \pi ab = \pi a^2\sqrt{1 - \varepsilon^2}. \tag{4.9}$$

If the eccentricity is zero, this reduces to the usual expression for the area of a circle of radius $a$.

## 4.3 Area as a Vector Cross-Product, and Kepler's Second Law

Consider a triangle formed from two vectors $\boldsymbol{A}$ and $\boldsymbol{B}$: $\boldsymbol{A}$ forms the base, and $\boldsymbol{B}$ is oriented at an angle $\psi$ to the direction of $A$ as shown in Figure 4.3. We wish to get an expression for the shaded area enclosed by the triangle in terms of the magnitudes $A$ and $B$ and the angle $\psi$.

If the right side of the triangle were at 90° to the base, the area would be just one-half of the product of the base and the height. The shaded area is then the area of the large right-angled triangle less that of the small right-angled triangle at the right end formed by the dashed lines:

$$A_\Delta = \frac{1}{2}(A + B\cos\psi)(B\sin\psi) - \frac{1}{2}(B\cos\psi)(B\sin\psi) = \frac{1}{2}AB\sin\psi. \quad (4.10)$$

Here is the point: $AB\sin\psi$ is precisely the magnitude of the cross-product $\boldsymbol{A} \times \boldsymbol{B}$, so we can write the area as

$$A_\Delta = \frac{1}{2}\,|\boldsymbol{A} \times \boldsymbol{B}|. \quad (4.11)$$

But what has this got to do with the physics of orbits? Imagine now that $\boldsymbol{A}$ is the position vector of a planet of reduced mass $\mu$ as measured from the Sun, which is at the tail of $\boldsymbol{A}$, while $\boldsymbol{B}$ points in the direction of the velocity vector $\boldsymbol{v}$ of the planet at the moment shown. If some increment of time $\Delta t$ elapses, then the length of $\boldsymbol{B}$ will be $v\Delta t$. This conception lets us write the area of the triangle swept out by the position vector over time $\Delta t$ in terms of the orbital angular momentum $L$ of the planet, by working through the linear momentum $\boldsymbol{p}$:

$$A_\Delta = \frac{1}{2}\,|\boldsymbol{r} \times \boldsymbol{v}\Delta t| = \frac{1}{2}\left|\boldsymbol{r} \times \frac{\boldsymbol{p}}{\mu}\Delta t\right| = \frac{\Delta t}{2\mu}\,|\boldsymbol{r} \times \boldsymbol{p}| = \frac{\Delta t}{2\mu}\,|\boldsymbol{L}| = \frac{\Delta t\, L}{2\mu} \quad (4.12)$$

Now, we already know that for an object orbiting under the action of a central force, angular momentum is conserved. *For equal time increments, the right side of equation (4.12) is a constant. This means that the position vector from the Sun to the planet will sweep out equal areas in equal times: Kepler's second law!*

Kepler deduced his second law by plotting planetary positions as a function of time; we now regard it as a manifestation of the conservation of angular momentum through the action of a central force.

Now put Equations (4.9) and (4.12) together in the sense of setting $\Delta t$ to be the full orbital period $T$ of the planet. During this time the planet's trajectory will encompass the full area of its elliptical orbit, so we can write

$$\pi a^2\sqrt{1 - \varepsilon^2} = \frac{LT}{2\mu} \Rightarrow L = \frac{2\mu\pi a^2}{T}\sqrt{1 - \varepsilon^2}. \quad (4.13)$$

In the following chapter we will derive an explicit expression for $L$ in terms of $a$, $\varepsilon$, $G$, and the masses involved; this will allow us to use Equation (4.13) to deduce Kepler's third law.

**Exercise:** In polar coordinates, the position vector is given by $\boldsymbol{r} = r\,\hat{\boldsymbol{r}}$, and an element of arc-length is given by $\boldsymbol{ds} = dr\,\hat{\boldsymbol{r}} + r\,d\phi\,\hat{\boldsymbol{\phi}}$. From the above analysis, an element of area is then $(\boldsymbol{r} \times \boldsymbol{ds})/2$. Integrate this around one-half of an ellipse and multiply by two; you should reproduce Equation (4.9).

## 4.4 How Did Kepler Plot the Orbits?

How did Johannes Kepler turn data on the positions of planets in the sky into maps of their elliptical orbits? This work, a triumph of basic geometry, is described in this section.

Historically, planets were divided into two classes: inferior and superior. Inferior planets are those whose orbits lie within that of Earth's: Mercury and Venus. The orbits of superior planets lie beyond that of the Earth; in Kepler's time, these were Mars, Jupiter, and Saturn.

Determining a planet's orbit is complicated by the fact that Earth itself also orbits the Sun. However, Earth's orbital eccentricity is small, so for the present purpose we can regard it as being a circle of radius one AU (recall that 1 AU is the average Earth–Sun distance). How the eccentricity and orientation of Earth's orbit can be deduced are described in Section 8.7.

First, consider the speed of a planet in a circular orbit around the Sun. Equating the gravitational and centripetal forces gives $GMm/r^2 = mv^2/r$, so the speed is $v = \sqrt{GM/r}$; *more distant planets must travel more slowly than closer ones.* Copernicus had inferred this from the available observational data.

The situation for an inferior planet, say Venus, is particularly straightforward. Consult Figure 4.4. There will be moments in Venus' orbit where it appears at a maximum angular separation from the Sun; astronomers know this situation as "maximum elongation." As seen from Earth, the line of sight to Venus at this moment must run tangent to its orbit, and so the Sun–Venus distance $d_{\mathrm{Venus}}$ must make a right angle to the Earth–Venus line of sight. If the angular separation of Venus from the Sun at this moment is $\beta$ degrees, then the definition of a sine as "opposite over hypotenuse" tells us that $d_{\mathrm{Venus}} = \sin\beta$ AU since the hypotenuse of the right-angled triangle, the Earth–Sun distance, is one AU. By tracking successive maximum elongations, the shape of Venus' orbit can be plotted.

The situation for a superior planet, say Mars, is somewhat more involved as it requires coordinating observations taken at two different times. Consult Figure 4.5. There will be moments when Mars appears in the sky in a direction exactly opposite

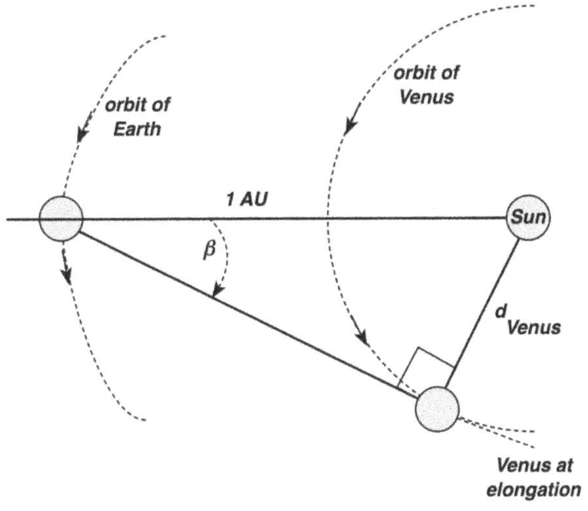


**Figure 4.4.** Determining the heliocentric distance of an inferior planet.

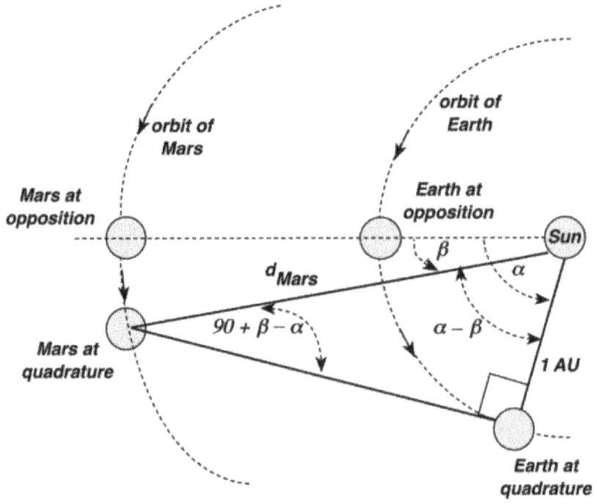


**Figure 4.5.** Determining the heliocentric distance of a superior planet.

to that of the Sun; the technical name for this is in fact "opposition." In practical terms, Mars will rise just as the Sun sets, or vice-versa.

With its closer orbit, Earth moves faster, and so a few weeks later Mars will appear exactly 90° from the Sun; this is known as the planet being "in quadrature." If the time between opposition and quadrature is known and the orbital periods of Earth and the planet are known (which they were; this is described below), then the angles $\alpha$ and $\beta$ through which Earth and Mars have orbited can be determined. The angle between the directions to the Sun and the Earth as measured from Mars is then $90° + \beta - \alpha$, and right-angle trigonometry tells us the Sun–Mars distance at the moment of quadrature: $d_{\mathrm{Mars}} = 1/\sin(90° + \beta - \alpha)$ AU. Successive opposition-to-quadrature configurations from Tycho's data allowed Kepler to map out Mars' orbit. Kepler expressed all distances in AUs and periods in years, but the physical laws involved are indifferent to the system of units used; the size of the AU in meters was not determined with any accuracy until a transit of Venus across the face of the Sun was timed in the 1700s.

How were the orbital periods known? The orbital period $T$ in Kepler's laws, e.g., Equation (4.13), is known as a "sidereal" period; this is the orbital period that would be reported by a "fixed" observer outside the solar system. In astronomical parlance, "sidereal" means "with respect to the distant stars." But the Earth itself is in orbit around the Sun, so the sidereal period is not directly observable. What is observable, however, is the time between successive conjunctions or oppositions of a planet with the Sun as described above. This is known as the "synodic" period. The relationship between sidereal and synodic periods was first deduced by Copernicus, who posited a heliocentric model for the solar system a generation before Kepler.

First consider the case of a superior planet such as Mars. Times in this argument are measured in Earth years. Imagine Earth and Mars as runners in a race around a circular track, with Earth running the inside lane and going a little faster than Mars. They start in opposition, lined up with the Sun. Earth completes one orbit, 360°, per year.

They start simultaneously, and Earth is equipped with a clock which measures elapsed time in years. As the race progresses, Earth will pull ahead of and eventually lap Mars (the next opposition), perhaps after both have run several laps. The time at which this happens will be one Martian synodic period; call this $S$ years. No matter how many laps each has run, Earth will have covered one more, or exactly 360° more than Mars when this occurs. Thus, Earth will have run through $S$ 360°, and if the *sideral* period of Mars is $T_{\text{Sup}}$ years (now using a subscript for a general superior planet), it will cover $360°/T_{\text{Sup}}$ per year, or $S\ 360°/T_{\text{Sup}}$ after the synodic period. This means that we must have $S\ 360° - S\ 360°/T_{\text{Sup}} = 360°$, or, on solving for $T_{\text{Sup}}$,

$$T_{\text{Sup}} = \frac{S}{S-1}. \tag{4.14}$$

For an inferior planet the logic is similar, except that the inferior planet covers the extra 360°. This leads to

$$T_{\text{Inf}} = \frac{S}{S+1}. \tag{4.15}$$

As an example, the synodic period of Venus is about 583.92 days = 1.599 years (1 year = 365.25 days). This gives $T_{\text{Venus}} = 1.599/2.599 = 0.615$ years, or about 224.7 days, about seven and a half months.

Table P.1 in the Preface lists synodic periods for the planets; check that these concur with the sidereal periods.

Clearly, this analysis will not hold for elliptical orbits, where the planets' speeds continuously vary; Copernicus was fortunate in that planetary orbits are not extremely eccentric. Later, Kepler was fortunate in that Tycho's post-Copernicus data on planetary positions were sufficiently accurate that he could tease out the eccentricities.

Spend a few minutes reflecting on what Tycho, Copernicus, and Kepler accomplished: with clocks, naked-eye observations with sextant-like instruments, and what is now high-school geometry, they deduced the layout of the solar system, setting the stage for Newton to derive orbital physics. This understanding of celestial mechanics is one of the most staggering intellectual achievements of humanity.

## 4.5 The Optical Theorem for Ellipses

This section can be considered optional in that it does not deal with orbital mechanics but rather an interesting geometric property of ellipses, the so-called "optical theorem" or "reflection theorem." This theorem states that if the inside of an ellipse (or, more generally, of an ellipse rotated about its major axis to create a three-dimensional ellipsoid) is reflective, then a light beam emitted from one focus must pass through the other focus. Life inside a reflective ellipsoid with a light source at one focus would be interesting! In acoustics, an analog of this effect known as a "whispering gallery" involves an elliptical enclosure beneath a dome or vault where conversation originating at one focus can be heard at the other.

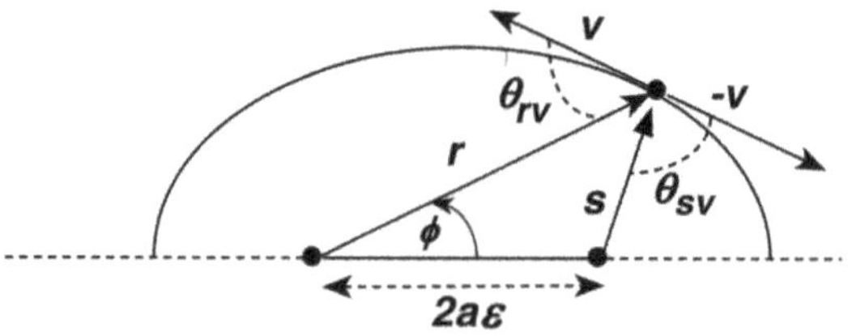


**Figure 4.6.** Top half of Figure 4.1 with the distances $r$ and $s$ of the orbiting planet from the foci rendered as vectors, $\boldsymbol{r}$ and $\boldsymbol{s}$.

This can be proven with a fairly straightforward vector analysis. Figure 4.6 shows the top half of Figure 4.1, with the distances $r$ and $s$ of the orbiting planet from the foci rendered as vectors, $\boldsymbol{r}$ and $\boldsymbol{s}$. The velocity vector $\boldsymbol{v}$ lies tangent to the ellipse; the negative of this, $-\boldsymbol{v}$, is also shown. The angle $\theta_{rv}$ lies between the vectors $\boldsymbol{v}$ and $-\boldsymbol{r}$, and the angle $\theta_{sv}$ lies between $-\boldsymbol{v}$ and $-\boldsymbol{s}$. If a light ray leaves the left focus and follows vector $\boldsymbol{r}$, its "angle of incidence" will be $\theta_{rv}$. If the "angle of reflection" $\theta_{sv}$ proves to be the same as $\theta_{rv}$, then the beam must follow the vector $-\boldsymbol{s}$ back through the right focus; this is the essence of the optical theorem. Our problem is to prove that $\theta_{sv} = \theta_{rv}$.

To begin, note that the $\boldsymbol{s}$ and $\boldsymbol{r}$ are related as

$$\boldsymbol{r} = (2a\varepsilon)\hat{\boldsymbol{x}} + \boldsymbol{s}. \tag{4.16}$$

Since $a$, $\varepsilon$, and $\hat{\boldsymbol{x}}$ are constants once the dimensions of the ellipse have been fixed, we must have

$$\frac{d\boldsymbol{r}}{dt} = \frac{d\boldsymbol{s}}{dt} = \boldsymbol{v}. \tag{4.17}$$

Also, from the definition of an ellipse, Equation (4.1), the rates of change of the magnitudes $r$ and $s$ relate as

$$s = 2a - r \Rightarrow \frac{ds}{dt} = -\frac{dr}{dt}. \tag{4.18}$$

The negative sign here indicates that as $r$ or $s$ increases, the other must decrease in order to conserve their sum.

The proof that $\theta_{sv} = \theta_{rv}$ makes use of the definition of a vector dot product, $\boldsymbol{A} \cdot \boldsymbol{B} = AB\cos\theta_{AB}$, where $A$ and $B$ are the magnitudes of the vectors $\boldsymbol{A}$ and $\boldsymbol{B}$ and $\theta_{AB}$ is the angle between them.

Consider the dot product of $-\boldsymbol{r}$ and $\boldsymbol{v}$:

$$-\boldsymbol{r} \cdot \boldsymbol{v} = rv\cos\theta_{rv} \Rightarrow v\cos\theta_{rv} = -\frac{1}{r}(\boldsymbol{r} \cdot \boldsymbol{v}) = -\frac{1}{r}\left(\boldsymbol{r} \cdot \frac{d\boldsymbol{r}}{dt}\right). \tag{4.19}$$

The term in brackets here can be reformulated as

$$\left(\boldsymbol{r}\cdot\frac{d\boldsymbol{r}}{dt}\right)=\frac{1}{2r}\frac{d}{dt}(\boldsymbol{r}\cdot\boldsymbol{r})=\frac{1}{2}\frac{d}{dt}(r^2)=\frac{1}{2}\left(2r\frac{dr}{dt}\right)=\left(r\frac{dr}{dt}\right). \tag{4.20}$$

This manipulation is similar to that utilized in the analysis of the work-energy theorem in Section 3.2. With this, (4.19) becomes

$$v\cos\theta_{rv}=-\frac{dr}{dt}. \tag{4.21}$$

Now perform the same analysis with the dot product of $-\boldsymbol{s}$ and $-\boldsymbol{v}$, using (4.17) to replace $\boldsymbol{v}$ with $d\boldsymbol{s}/dt$:

$$(-\boldsymbol{s})\cdot(-\boldsymbol{v})=sv\cos\theta_{sv}\Rightarrow v\cos\theta_{sv}=\frac{1}{s}(\boldsymbol{s}\cdot\boldsymbol{v})=\frac{1}{s}\left(\boldsymbol{s}\cdot\frac{d\boldsymbol{s}}{dt}\right), \tag{4.22}$$

which becomes, in analogy to (4.21),

$$v\cos\theta_{sv}=\frac{1}{2s}\frac{d}{dt}(\boldsymbol{s}\cdot\boldsymbol{s})=\frac{1}{2s}\frac{d}{dt}(s^2)=\frac{1}{2s}\left(2s\frac{ds}{dt}\right)=-\frac{dr}{dt}, \tag{4.23}$$

where the last step follows from (4.18). Equations (4.21) and (4.23) show that $\theta_{sv}=\theta_{rv}$, verifying the optical theorem. To actually determine this angle, you need expressions for $dr/dt$ and $v$ as functions of the polar angle $\phi$; for these, see Chapter 5.

# Chapter 5

## Elliptical Orbits and the Inverse-Square Law: Geometry Meets Physics

This chapter is the heart of this book. With the results of the preceding chapters, we are now prepared to understand how Newton's inverse-square law of gravitation leads to Kepler's elliptical orbits.

There are two ways to approach this issue, and both are covered in this chapter. First, we can start by assuming that the orbits are elliptical as Kepler had shown empirically, and work backwards to the presumed-radial force law by seeing what happens when we substitute the equation for an ellipse into Equation (3.53). This is done in Section 5.1, and some of the consequences for energy, angular momentum, and velocity are explored in Section 5.2. Section 5.3 looks at the second method: assuming that Newton's inverse-square law is correct, and directly integrating Equation (3.42) to deduce the form of the resulting orbit. To be sure, this approach is conditioned by knowing Kepler's empirical answer, but this does not detract from the correctness of the solution. Section 5.4 develops a proof of Kepler's third law, and Section 5.5 establishes an equation which explicitly links time and the angular position $\phi$ of our orbiting reduced mass. Section 5.6 offers an example of analyzing the trajectory of an Earth-orbiting spy satellite. The remaining sections examine some optional but interesting issues. In advanced dynamics and quantum mechanics classes, you may encounter a construct known as the *Laplace–Runge–Lenz* vector. Applied to an inverse-square force, this yields another conserved quantity for an elliptical orbit; this is done in Section 5.7. Section 5.8 takes a brief look at how Kepler's third law would behave for central forces of the more general magnitude $F(r) \propto r^n$, where $n$ is presumed to be an integer; this connects to Newton's shell-point equivalency theorem of Section 3.7. Section 5.9 considers a graphical way of examining orbits via a method known as the *effective potential.* Section 5.10 examines basic radial perturbation theory for circular orbits, and Section 5.11 looks at the issue of escape velocity.

doi:10.1088/2514-3433/acb430ch5

## 5.1 Proof by Assuming an Elliptical Orbit: Angular Momentum

To begin this section, we open by reiterating Equation (3.53) for the behavior of a central force in terms of the radial distance $r$ and azimuthal angle $\phi$, and Equation (4.3) for the shape of an ellipse in terms of the same coordinates:

$$F(r) = -\left(\frac{L^2}{\mu r^2}\right)\left\{\frac{1}{r^2}\left(\frac{d^2 r}{d\phi^2}\right) - \frac{2}{r^3}\left(\frac{dr}{d\phi}\right)^2 - \frac{1}{r}\right\}, \tag{5.1}$$

and

$$r = \frac{a(1 - \varepsilon^2)}{1 - \varepsilon \cos \phi}. \tag{5.2}$$

Remember that the angular momentum $L$ can be treated as a constant for motion under a central force. The approach here is to substitute Equation (5.2) into Equation (5.1) and see what emerges for $F(r)$. If the resulting expression is a function only of $r$, that is, if there are no un-removable factors of $\phi$ left over, then we will have shown that an elliptical orbit does correspond to a central force, and we will learn how the force depends on $r$.

For convenience, define the numerator of Equation (5.2) as

$$\beta = a(1 - \varepsilon^2). \tag{5.3}$$

We need the first and second derivatives of $r(\phi)$. The first derivative is

$$\frac{dr}{d\phi} = -\frac{\beta\, \varepsilon \sin \phi}{(1 - \varepsilon \cos \phi)^2} = -\frac{r^2 \varepsilon}{\beta} \sin \phi. \tag{5.4}$$

The second derivative is

$$\frac{d^2 r}{d\phi^2} = -\frac{\varepsilon}{\beta}\left[2r\left(\frac{dr}{d\phi}\right)\sin \phi + r^2 \cos \phi\right]. \tag{5.5}$$

Substitute Equation (5.4) for $(dr/d\phi)$ into Equation (5.5), and extract a factor of $r^2$ to give

$$\frac{d^2 r}{d\phi^2} = \frac{\varepsilon r^2}{\beta}\left[\frac{2r\, \varepsilon \sin^2 \phi}{\beta} - \cos \phi\right]. \tag{5.6}$$

Now substitute Equations (5.4) and (5.6) into Equation (5.1). After a few lines of algebra, you should find

$$F(r) = \left(\frac{L^2}{\mu r^2}\right)\left\{\frac{\varepsilon \cos \phi}{\beta} + \frac{1}{r}\right\}. \tag{5.7}$$

On replacing $1/r$ with Equations (5.2) and (5.3), this reduces to

$$F(r) = \frac{L^2}{\mu\,\beta\,r^2} = \frac{L^2}{\mu\,a(1-\varepsilon^2)\,r^2}. \tag{5.8}$$

*The result is a purely-radial inverse-square force!* If we marry this result with Newton's hypothesized force law $F(r) = GMm/r^2$, we find

$$\frac{GMm}{r^2} = \frac{L^2}{\mu\,a(1-\varepsilon^2)\,r^2}, \tag{5.9}$$

or

$$L^2 = GMm\mu a(1-\varepsilon^2) = -\kappa\mu a(1-\varepsilon^2) = \frac{G(Mm)^2}{(M+m)}a(1-\varepsilon^2). \tag{5.10}$$

*Physical interpretation: The size and shape of the elliptical orbit in combination with the masses involved dictate the angular momentum of the system. This is where the geometry of Kepler's ellipses meets the physics of Newton's gravitational law.*

**Exercise:** Verify that the units of Equation (5.10) are correct.

## 5.2 Velocity, the *Vis-viva* Equation, and Energy

The expression for angular momentum derived in the preceding section is the gateway to much of the physics of elliptical orbits. We begin this section by deriving expressions for the speed of the planet at any radial or angular position in its orbit, and then one for the total energy of the system.

From Equation (2.9), the square of the speed is given by

$$v^2 = (\dot{r})^2 + (r\,\dot{\phi})^2. \tag{5.11}$$

$\dot{r}$ and $r\,\dot{\phi}$ are, respectively, the radial and tangential speeds. For the radial speed, differentiate the equation of the ellipse, (5.2), with respect to time,

$$\frac{dr}{dt} = -\frac{a(1-\varepsilon^2)\,\varepsilon\sin\phi}{(1-\varepsilon\cos\phi)^2}\left(\frac{d\phi}{dt}\right) = -\frac{r^2\varepsilon}{a(1-\varepsilon^2)}\sin\phi\left(\frac{d\phi}{dt}\right). \tag{5.12}$$

For $(d\phi/dt)$, invoke Equation (3.35) (again!) and then use Equation (5.10) for $L$:

$$\begin{aligned} v_{\text{rad}} &= \frac{dr}{dt} = -\frac{r^2\varepsilon}{a(1-\varepsilon^2)}\sin\phi\left(\frac{L}{\mu r^2}\right) \\ &= -\frac{L\varepsilon}{a(1-\varepsilon^2)\mu}\sin\phi = -\varepsilon\sin\phi\sqrt{\frac{G(M+m)}{a(1-\varepsilon^2)}}. \end{aligned} \tag{5.13}$$

The tangential speed similarly reduces to

$$v_{\text{tang}} = r\,\dot{\phi} = \sqrt{\frac{G(M+m)}{a(1-\varepsilon^2)}}(1-\varepsilon\cos\phi). \tag{5.14}$$

The overall speed is the square root of the sum of the squares of the components:

$$v = \sqrt{\frac{G(M+m)}{a(1-\varepsilon^2)}(1+\varepsilon^2 - 2\varepsilon\cos\phi)}. \tag{5.15}$$

At aphelion ($\phi = 0°$ in Figure 4.1), Equations (5.2) and (5.15) give

$$(r, v)_{\text{aphelion}} = \left[a(1+\varepsilon), \quad \sqrt{\frac{G(M+m)}{a}\frac{(1-\varepsilon)}{(1+\varepsilon)}}\right], \tag{5.16}$$

whereas at perihelion ($\phi = 180°$)

$$(r, v)_{\text{perihelion}} = \left[a(1-\varepsilon), \quad \sqrt{\frac{G(M+m)}{a}\frac{(1+\varepsilon)}{(1-\varepsilon)}}\right]. \tag{5.17}$$

Notice that $v_{\text{perihelion}} > v_{\text{aphelion}}$: the planet moves faster when it is closer to the Sun.

Another version of Equation (5.15) can be had by eliminating $\varepsilon\cos\phi$ through the equation of the ellipse. This gives

$$v^2 = G(M+m)\left(\frac{2}{r} - \frac{1}{a}\right). \tag{5.18}$$

This expression is known as the *vis-viva* equation, which is Latin for "living force." Equation (5.15) gives $v(\phi)$, whereas this result gives $v(r)$.

We come now to an *extremely important* result: an expression for the total energy $E$ of the system, $E = \mu v^2/2 + V(r)$. With Equation (5.18) and (3.23), this is

$$E = \frac{\mu}{2}G(M+m)\left(\frac{2}{r} - \frac{1}{a}\right) - \frac{GMm}{r} = -\frac{GMm}{2a}. \tag{5.19}$$

*Physical interpretation: The size (semimajor axis) of the elliptical orbit dictates the total energy of the system. The total energy must be negative for "a" to be positive; an elliptical orbit is a "bound energy" state. We can then think of the total angular momentum as being a reflection of the shape of the orbit through the eccentricity in Equation (5.10). There are two conserved quantities in the physics, and two parameters are needed to describe an ellipse.*

**Exercise:** Look at Equation (5.10). For orbits of the same energy, that is, ones with the same value of $a$, one of greater eccentricity will have less angular momentum. Develop a physical argument as to why a "skinny" orbit whose semimajor axis is the same as the radius of a circular one will have less angular momentum.

## 5.3 Proof of Elliptical Orbits by Direct Integration

This section offers an alternate proof of the fact that an inverse-square force law gives rise to elliptical orbits by directly integrating the angular trajectory equation,

(3.42). I begin by re-writing this as an indefinite integral, with the constant factors of $L$ in the numerator and $2/\mu$ in the denominator extracted:

$$\phi = \frac{L}{\sqrt{2\mu}} \int \frac{dr}{r^2\sqrt{E - V(r) - L^2/2\mu r^2}}. \tag{5.20}$$

The potential $V(r)$ is that of Equation (3.23):

$$V(r) = -\frac{GMm}{r} = +\frac{\kappa}{r}. \tag{5.21}$$

Substitute Equation (5.21) into (5.20). Also extract one factor of $r$ from the $r^2$ in front of the radical, square it, and move it inside the radical so that no factors of $r$ remain in the denominators of the terms within:

$$\phi = \frac{L}{\sqrt{2\mu}} \int \frac{dr}{r\sqrt{Er^2 - \kappa r - L^2/2\mu}}. \tag{5.22}$$

This integral is of the standard form [see (1.15)]

$$\int \frac{dx}{x\sqrt{cx^2 + bx + d}} = \frac{1}{\sqrt{-d}} \sin^{-1}\left[\frac{bx + 2d}{|x|\sqrt{-q}}\right], \tag{5.23}$$

where

$$q = 4\,cd - b^2. \tag{5.24}$$

We have

$$\left.\begin{aligned} d &= -L^2/2\mu \\ b &= -\kappa \\ c &= E \end{aligned}\right\}. \tag{5.25}$$

The factor of $\sqrt{-d}$ reduces to $L/\sqrt{2\mu}$, which cancels that in front of the integral in Equation (5.22). Hence

$$\phi = \sin^{-1}\left[-\frac{(\kappr + L^2/\mu)}{r\sqrt{-q}}\right] + C, \tag{5.26}$$

where $C$ is a constant of integration

This certainly looks nothing like the equation of an ellipse—so far. As for the factor of $\sqrt{-q}$, I define this as $Q$:

$$Q = \sqrt{-q} = \sqrt{\frac{2L^2E}{\mu} + \kappa^2}. \tag{5.27}$$

Two properties of the inverse-sine function are useful here. These are

$$\sin^{-1}(-x) = -\sin^{-1}(x), \tag{5.28}$$

and

$$\sin^{-1}(x) = \frac{\pi}{2} - \cos^{-1}(x). \tag{5.29}$$

These let us write $\sin^{-1}(-x) = \cos^{-1}(x) - \pi/2$. Invoke this in (5.26). Then with (5.27) and a few steps of algebra, $r$ can be extracted as

$$r = \frac{(L^2/\mu)}{Q\cos(\phi - C + \pi/2) - \kappa}. \tag{5.30}$$

Now return to $Q$ of Equation (5.27). This next step may seem strange, but, as argued below, is entirely legitimate. Adopt Equations (5.10) and (5.19) from the preceding section for $L$ and $E$:

$$L^2 = GMm\mu a(1 - \varepsilon^2), \tag{5.31}$$

and

$$E = -\frac{GMm}{2a}. \tag{5.32}$$

This looks as if we are "assuming the answer," but the intent here is to use (5.31) and (5.32) to define two new variables, $a$ and $\varepsilon$, to replace $L$ and $E$. This is what was meant by the comment at the beginning of this chapter that this calculation is conditioned by knowing the answer, but it is an entirely legitimate process.

With these definitions, you should find that

$$Q = (GMm)\varepsilon. \tag{5.33}$$

Insert this result and the definition of $\kappa$ into Equation (5.30). A common factor of $GMm$ appears in the denominator, which can be extracted to give

$$r = \frac{[L^2/\mu GMm]}{\varepsilon\cos(\phi - C + \pi/2) + 1}. \tag{5.34}$$

With Equation (5.31), the numerator here reduces to $a(1 - \varepsilon^2)$:

$$r = \frac{a(1 - \varepsilon^2)}{\varepsilon\cos(\phi - C + \pi/2) + 1}, \tag{5.35}$$

an expression which is beginning to look somewhat like the equation of an ellipse.

Now we have to deal with the constant of integration $C$. This is set by deciding upon the value of $r$ for some specific value of $\phi$. We have complete liberty to set this initial condition as we please, so long as it is physically realistic, i.e., $r \neq 0$. I set this as

$$r = a(1 + \varepsilon) \quad \text{at} \quad \phi = 0. \tag{5.36}$$

On substituting this into Equation (5.35), what remains is a condition on $C$:

$$\cos\left(\frac{\pi}{2} - C\right) = -1. \tag{5.37}$$

This can only be satisfied if $C = -\pi/2$, which renders Equation (5.35) as

$$r = \frac{a(1 - \varepsilon^2)}{\varepsilon \cos(\phi + \pi) + 1}. \tag{5.38}$$

However, $\cos(\phi + \pi) = -\cos\phi$, so we have

$$r = \frac{a(1 - \varepsilon^2)}{1 - \varepsilon \cos\phi}, \tag{5.39}$$

completing the somewhat contrived proof.

## 5.4 Kepler's Third Law

This section develops a proof of Kepler's third law: that the square of the orbital period $T$ of a planet is proportional to the cube of its semimajor axis. This analysis of such a significant result will be so short as to seem almost anti-climactic.

This proof makes use of the development in Section 4.3 of a relationship between the angular momentum of the orbit and the period, Equation (4.13), here written squared:

$$\pi^2 a^4(1 - \varepsilon^2) = \frac{L^2 T^2}{4\mu^2}. \tag{5.40}$$

But we now have an explicit expression for $L^2$ from Equation (5.10):

$$L^2 = GMm\mu a(1 - \varepsilon^2). \tag{5.41}$$

Combining these gives

$$\pi^2 a^4(1 - \varepsilon^2) = \frac{T^2}{4\mu^2} GMm\mu a(1 - \varepsilon^2) \;\Rightarrow\; T^2 = \left[\frac{4\pi^2}{G(M + m)}\right] a^3, \tag{5.42}$$

which is Kepler's third law.

The $T^2$ versus $a^3$ relationship for the traditional nine planets of the solar system is shown in Figure 5.1; to keep the inner planets from clumping together in the lower-left part of the plot, the scales are logarithmic. For a power law relationship of the form $T = KT^{3/2}$ ($K$ = constant), taking the logarithm of both sides gives $\log(T) = \log K + (3/2)\log(a)$. A graph of $\log(T)$ versus $\log(a)$ should then be a straight line of slope 3/2, as can be seen to be the case. With periods and years and semimajor axes in Astronomical Units (AUs), the constant $K$ evaluates to exactly $K = 1$, so $\log(K) = 0$: the intercept is zero, as can also be seen to be the case. To see how you can show that $K = 1$ for these units, consult Section 8.6.

In Section 5.8, we will revisit Kepler's third law to see how it works out for power-law central forces with powers other than 2.

**Exercise:** Pluto orbits the Sun with semimajor axis $a$ = 39.48 AU. See Table P.1. What are Pluto's period, maximum and minimum distances from the Sun, and maximum and minimum orbital speeds? Answer: 248 years, 49.3 AU, 29.6 AU, 6.1 km s$^{-1}$, 3.7 km s$^{-1}$.

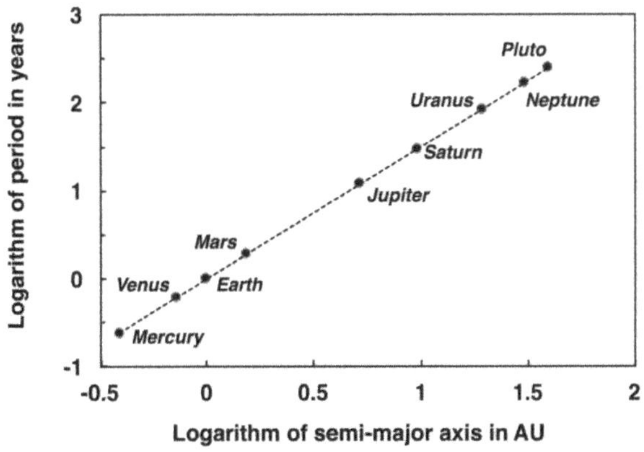


**Figure 5.1.** Period versus semimajor axis relationship for the traditional nine planets of the solar system. Pluto is now considered to be a dwarf planet. That the points lie along a line of slope 3/2 confirms Kepler's third law.

**Exercise:** Combine Equations (5.10) and (5.42) to show that the angular momentum can be written as

$$L = \frac{GT(Mm)}{2\pi a}\sqrt{1 - \varepsilon^2}.$$

## 5.5 The Time–Angle equation

This section develops an explicit relationship linking time and the angular position $\phi$ of an orbiting planet.

This proof begins with the workhorse of many of our derivations, Equation (3.35), written explicitly for an elliptical orbit and invoking Equation (5.10) for the angular momentum:

$$\begin{aligned}\frac{d\phi}{dt} &= \frac{L}{\mu\, r^2} \\ &= \frac{\sqrt{GMm\mu a(1 - \varepsilon^2)}}{\mu}\frac{(1 - \varepsilon\cos\phi)^2}{a^2(1 - \varepsilon^2)^2} \\ &= \frac{\sqrt{G(M + m)}}{a^{3/2}(1 - \varepsilon^2)^{3/2}}(1 - \varepsilon\cos\phi)^2.\end{aligned} \tag{5.43}$$

The leading factor can be expressed in terms of the orbital period via Kepler's third law, Equation (5.42):

$$\frac{d\phi}{dt} = \frac{2\pi}{T(1 - \varepsilon^2)^{3/2}}(1 - \varepsilon\cos\phi)^2. \tag{5.44}$$

Separating variables gives

$$\frac{dt}{T} = \frac{(1-\varepsilon^2)^{3/2}}{2\pi}\frac{d\phi}{(1-\varepsilon\cos\phi)^2}. \tag{5.45}$$

Integrate this expression, assuming that $\phi = 0$ at $t = 0$, that is, that the motion starts at perihelion at time-zero. This is not a mandatory choice, but makes the final result more compact; I write a subscript on the left side as a reminder of this choice:

$$\left(\frac{t}{T}\right)_{0\to\phi} = \frac{(1-\varepsilon^2)^{3/2}}{2\pi}\int_0^{\phi}\frac{d\phi}{(1-\varepsilon\cos\phi)^2}. \tag{5.46}$$

The integral appearing here is that of Equations (1.13) and (1.14), with $c = 1$ and $d = -\varepsilon$:

$$\left(\frac{t}{T}\right)_{0\to\phi} = \frac{(1-\varepsilon^2)^{3/2}}{2\pi} \times\left\{\frac{1}{(\varepsilon^2-1)}\left[\frac{-\varepsilon\sin\phi}{(1-\varepsilon\cos\phi)} - \frac{2}{\sqrt{1-\varepsilon^2}}\tan^{-1}\left[\sqrt{\frac{1+\varepsilon}{1-\varepsilon}}\tan\left(\frac{\phi}{2}\right)\right]\right]\right\}_0^{\phi}.$$

This simplifies to

$$\left(\frac{t}{T}\right)_{0\to\phi} = \frac{\sqrt{1-\varepsilon^2}}{2\pi}\left[\frac{\varepsilon\sin\phi}{(1-\varepsilon\cos\phi)} + \frac{2}{\sqrt{1-\varepsilon^2}}\tan^{-1}\left[\sqrt{\frac{1+\varepsilon}{1-\varepsilon}}\tan\left(\frac{\phi}{2}\right)\right]\right]_0^{\phi}. \tag{5.47}$$

Both terms vanish at the lower limit:

$$\left(\frac{t}{T}\right)_{0\to\phi} = \frac{\sqrt{1-\varepsilon^2}}{2\pi}\left[\frac{\varepsilon\sin\phi}{(1-\varepsilon\cos\phi)} + \frac{2}{\sqrt{1-\varepsilon^2}}\tan^{-1}\left[\sqrt{\frac{1+\varepsilon}{1-\varepsilon}}\tan\left(\frac{\phi}{2}\right)\right]\right]. \tag{5.48}$$

This is the main result of this section: an expression for $t(\phi)$. To be sure, it is a mess. Particularly annoying is the fact that there is no way we can transform it to solve for position a function of time: $\phi(t)$. Over the centuries, mathematicians have devised numerous approximate ways of doing so, but nowadays with lightning-fast programs and spreadsheets it is but a few seconds work to solve for the value of $\phi$ that corresponds to some value of $t/T$. A greatly simplified version of Equation (5.48) will be developed in Chapter 6, but it too cannot be rearranged to give $\phi(t)$ in a closed form.

**Exercise:** Consider one-half of an orbit, $\phi = \pi$. Show that $t/T = 1/2$. The comments following Equation (4.8) may be helpful.

## 5.6 Example: An Earth-Orbiting Spy Satellite

Suppose that we wish to put a satellite of mass $m = 1000$ kg into a polar orbit around the Earth with a period of exactly six hours so that it has a distance of closest approach to the surface of 70 km, as shown in Figure 5.2. With this period, the satellite will pass over the same four points on the Earth's surface once each every 24 h. What are the necessary semimajor axis of the orbit, its eccentricity, the speed of the satellite at apogee and perigee, the angular speed $d\phi/dt$ of the satellite at apogee and perigee, and the fraction of each orbit that the satellite will spend with $90° \leqslant \phi \leqslant 270°$, i.e., close to the Earth? Take $G = 6.674 \times 10^{-11}\,\text{m}^3\,\text{kg}^{-1}\,\text{s}^{-2}$, mass of Earth $M_E = 5.972 \times 10^{24}$ kg, and radius of Earth $R_E = 6371$ km. The center of the Earth is the focus of the orbit, and the dashed horizontal line is the equator.

In writing out the calculations, units will be suppressed within the formulae for brevity. Be sure to convert kilometers to meters and hours to seconds.

The semimajor axis comes directly from Kepler's Third Law:

$$
\begin{aligned}
a &= \left[\frac{G(M_E + m)T^2}{4\pi^2}\right]^{1/3} \\
&= \left[\frac{(6.674 \times 10^{-11})(5.972 \times 10^{24})(21{,}600)^2}{4\pi^2}\right]^{1/3} \\
&= 1.676 \times 10^7 \text{ m},
\end{aligned}
$$

or about 2.63 Earth radii.

The eccentricity can be computed from the perigee distance and Equation (5.17); note that $r_{\text{perigee}}$ is not 70 km, but (6371 + 70) = 6441 km: $r$ is measured from the center of the Earth. Here we can leave the distances in km because the units cancel:

$$
\varepsilon = 1 - \frac{r_{\text{perigee}}}{a} = 1 - \frac{6441}{16{,}760} = 0.616,
$$

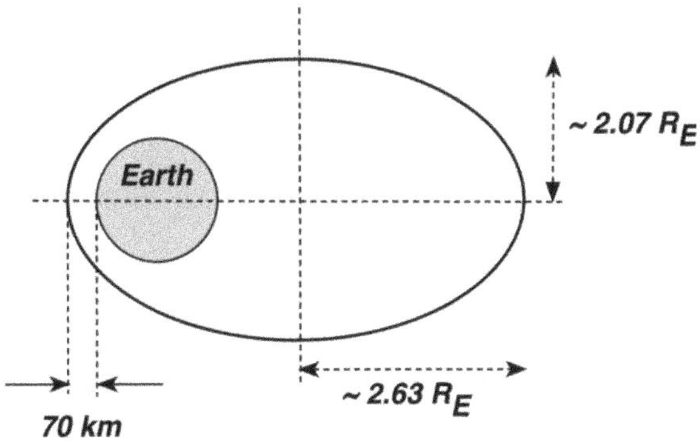


**Figure 5.2.** Orbit of a close-approaching Earth satellite with a period of six hours. This diagram is not to scale. $R_E$ = Earth radius.

a fairly eccentric orbit. The semiminor axis evaluates to $b = a\sqrt{1 - \varepsilon^2} \sim 2.07R_E$, and the satellite's maximum distance from the center of the Earth, $a(1 + \varepsilon)$, evaluates to $2.708 \times 10^7$ m, or about 4.25 Earth radii.

The speeds at apogee and perigee are

$$\begin{aligned} v_{\text{apogee}} &= \sqrt{\frac{G(M_E + m)}{a}\frac{(1 - \varepsilon)}{(1 + \varepsilon)}} \\ &= \sqrt{\frac{(6.674 \times 10^{-11})(5.972 \times 10^{24})}{(1.676 \times 10^7)}\frac{(0.384)}{(1.616)}} \\ &= 2380 \text{ m s}^{-1}, \end{aligned}$$

and

$$\begin{aligned} v_{\text{perigee}} &= \sqrt{\frac{G(M_E + m)}{a}\frac{(1 + \varepsilon)}{(1 - \varepsilon)}} \\ &= \sqrt{\frac{(6.674 \times 10^{-11})(5.972 \times 10^{24})}{(1.676 \times 10^7)}\frac{(1.616)}{(0.384)}} \\ &= 10{,}000 \text{ m s}^{-1}. \end{aligned}$$

That this latter figure works out so neatly to 10 km s$^{-1}$ is just an accident of the numbers. To compute $d\phi/dt$, use Equation (3.35), first computing the angular momentum $L$ from Equation (5.10). The mass of the satellite is so small compared to that of the Earth that the reduced mass is essentially exactly 1000 kg:

$$\begin{aligned} L &= \sqrt{GM_E m\mu a(1 - \varepsilon^2)} \\ &= \sqrt{(6.674 \times 10^{-11})(5.972 \times 10^{24})(1000)^2(1.676 \times 10^7)(0.621)} \\ &= 6.440 \times 10^{13} \text{ kg m}^2 \text{ s}^{-1}. \end{aligned}$$

At perigee:

$$\frac{d\phi}{dt} = \frac{L}{\mu r^2} = \frac{6.440 \times 10^{13}}{(1000)(6.441 \times 10^6)^2} = 1.552 \times 10^{-3} \text{ rad s}^{-1},$$

equivalent to about 320 degrees per hour! This is the angular speed that would be measured by an observed located at the center of the Earth. At apogee the angular speed is much smaller:

$$\frac{d\phi}{dt} = \frac{L}{\mu r^2} = \frac{6.440 \times 10^{13}}{(1000)(2.708 \times 10^7)^2} = 8.779 \times 10^{-5} \text{ rad s}^{-1},$$

or about 18 degrees per hour.

The satellite actually spends little time in close proximity to the Earth. It is left as an exercise for the reader to show by using Equation (5.48) that it spends only about 48 minutes between $\phi = 90°$ and 270°. This calculation is revisited in Chapter 6.

## 5.7 The Laplace–Runge–Lenz Vector

This section, which can be considered optional, deals with a mathematical construct known as the *Laplace–Runge–Lenz* (LRL) vector applied to central forces, specifically the inverse-square central force. You may encounter the LRL vector in more advanced dynamics and quantum mechanics classes. This vector is named after the French polymath Pierre-Simon de Laplace (1749–1827) whose name graces many areas of physics and mathematics, German mathematician/physicist Carl Runge (1856–1927), whose name may be familiar from the Runge–Kutta method of numerical integration, and German physicist Wilhelm Lenz (1888–1957). This is *not* the Lenz of Lenz's law of induction in electromagnetic theory, which is associated with Emil Lenz (1804–1865).

To begin, consider the time derivative of the vector cross product of the linear and angular momentum of a system:

$$\frac{d}{dt}(\boldsymbol{p} \times \boldsymbol{L}) = \frac{d\boldsymbol{p}}{dt} \times \boldsymbol{L} + \frac{d\boldsymbol{L}}{dt} \times \boldsymbol{p}. \tag{5.49}$$

Now, we know that $\boldsymbol{L}$ is conserved for a central force, so $d\boldsymbol{L}/dt = 0$. Newton's second law tells us that $d\boldsymbol{p}/dt = \boldsymbol{F}$, so we have

$$\frac{d}{dt}(\boldsymbol{p} \times \boldsymbol{L}) = \boldsymbol{F} \times \boldsymbol{L}. \tag{5.50}$$

Now consider the right side of this expression for a central force of the general form $\boldsymbol{F} = F(r)\hat{\boldsymbol{r}}$. What is had in mind is the force on our usual orbiting body of reduced mass $\mu$ toward a central "gravitating" body. Later we will put $F(r) = \kappa/r^2$ with $\kappa = -GMm$ per Equation (3.23), but for the moment the analysis will be left more general. Also write the angular momentum as $\boldsymbol{L} = \boldsymbol{r} \times \boldsymbol{p}$. This gives

$$\boldsymbol{F} \times \boldsymbol{L} = F(r)\hat{\boldsymbol{r}} \times (\boldsymbol{r} \times \boldsymbol{p}). \tag{5.51}$$

Now write $\hat{\boldsymbol{r}}$ as $\boldsymbol{r}/r$ and the linear momentum as $\boldsymbol{p} = \mu\boldsymbol{v} = \mu(d\boldsymbol{r}/dt)$, where $\boldsymbol{r}$ is the usual $M$-to-$m$ position vector. The scalar factor of $\mu$ can be extracted to give

$$\boldsymbol{F} \times \boldsymbol{L} = \frac{\mu F(r)}{r}\left[\boldsymbol{r} \times \left(\boldsymbol{r} \times \frac{d\boldsymbol{r}}{dt}\right)\right]. \tag{5.52}$$

The triple cross product can be simplified by utilizing an identity from vector analysis:

$$\boldsymbol{A} \times (\boldsymbol{B} \times \boldsymbol{C}) = \boldsymbol{B}(\boldsymbol{A} \cdot \boldsymbol{C}) - \boldsymbol{C}(\boldsymbol{A} \cdot \boldsymbol{B}). \tag{5.53}$$

Hence we have

$$\left[\boldsymbol{r} \times \left(\boldsymbol{r} \times \frac{d\boldsymbol{r}}{dt}\right)\right] = \boldsymbol{r}\left(\boldsymbol{r} \cdot \frac{d\boldsymbol{r}}{dt}\right) - \frac{d\boldsymbol{r}}{dt}(\boldsymbol{r} \cdot \boldsymbol{r}). \tag{5.54}$$

In the last term, we can set $(\boldsymbol{r} \cdot \boldsymbol{r}) = r^2$. The first term on the right side can also be simplified as we have done before:

$$\boldsymbol{r} \cdot \frac{d\boldsymbol{r}}{dt} = \frac{1}{2}\frac{d}{dt}(\boldsymbol{r} \cdot \boldsymbol{r}) = \frac{1}{2}\frac{d}{dt}(r^2) = r\frac{dr}{dt}. \tag{5.55}$$

Putting all this together gives

$$\frac{d}{dt}(\boldsymbol{p} \times \boldsymbol{L}) = \frac{\mu F(r)}{r}\left[\boldsymbol{r}\, r\frac{dr}{dt} - r^2\frac{d\boldsymbol{r}}{dt}\right]. \tag{5.56}$$

Be careful to keep straight which terms are scalars and which are vectors!

For a reason that will be clear in a moment, extract a factor of $r^3$ from within the square brackets of this expression:

$$\frac{d}{dt}(\boldsymbol{p} \times \boldsymbol{L}) = \mu F(r) r^2\left[\frac{\boldsymbol{r}}{r^2}\frac{dr}{dt} - \frac{1}{r}\frac{d\boldsymbol{r}}{dt}\right]. \tag{5.57}$$

The reason for this manipulation is that the square-bracketed term can be compacted as follows using the product rule (try it!):

$$\left[\frac{\boldsymbol{r}}{r^2}\frac{dr}{dt} - \frac{1}{r}\frac{d\boldsymbol{r}}{dt}\right] = -\frac{d}{dt}\left(\frac{\boldsymbol{r}}{r}\right). \tag{5.58}$$

With this, we can write Equation (5.57) as

$$\frac{d}{dt}(\boldsymbol{p} \times \boldsymbol{L}) = -\mu F(r) r^2\frac{d}{dt}\left(\frac{\boldsymbol{r}}{r}\right). \tag{5.59}$$

At this point, we need to specify $F(r)$ in order to proceed. To connect this development to Keplerian orbits, we set $F(r) = \kappa/r^2$ as described above. This causes the factor of $r^2$ on the right side of Equation (5.59) to cancel, leaving

$$\frac{d}{dt}(\boldsymbol{p} \times \boldsymbol{L}) = -\mu\kappa\frac{d}{dt}\left(\frac{\boldsymbol{r}}{r}\right),$$

which, because $\mu$ and $\kappa$ are constants, can be written as

$$\frac{d}{dt}\left[\boldsymbol{p} \times \boldsymbol{L} + \mu\kappa\left(\frac{\boldsymbol{r}}{r}\right)\right] = 0. \tag{5.60}$$

In the last term, $(\boldsymbol{r}/r)$ could be written as $\hat{\boldsymbol{r}}$. The physical conclusion here is that for an inverse-square central force, the time-derivative of the expression in square brackets is zero, *hence it must be a constant quantity*. This is the LRL vector, usually designated as $\boldsymbol{A}$:

$$\boldsymbol{A} = \boldsymbol{p} \times \boldsymbol{L} + \mu\kappa\left(\frac{\boldsymbol{r}}{r}\right). \tag{5.61}$$

To show that the equation for an ellipse emerges from this is not easy, but we can do a consistency check by seeing if what we know of an elliptical trajectory and positing $\kappa = -GMm$ does lead to constant components for $\boldsymbol{A}$. The algebra in doing so is a little messy, but it does work out. This most easily done with a mixture of polar coordinates and Cartesian unit vectors, as we can treat the latter as constants.

First we need an expression for $\boldsymbol{A}$ broken out into components. To begin, $\boldsymbol{r}/r$ is just the unit vector $\hat{\boldsymbol{r}}$ of Equation (1.3):

$$\hat{\boldsymbol{r}} = (\cos\phi)\hat{\boldsymbol{x}} + (\sin\phi)\hat{\boldsymbol{y}}. \tag{5.62}$$

Also from Equation (1.3),

$$\hat{\boldsymbol{\phi}} = -(\sin\phi)\hat{\boldsymbol{x}} + (\cos\phi)\hat{\boldsymbol{y}}. \tag{5.63}$$

The linear momentum $\boldsymbol{p}$ is given by $\boldsymbol{p} = \mu\boldsymbol{v}$; the velocity vector $\boldsymbol{v}$ is given by Equation (2.3), so with Equations (5.62) and (5.63) we get

$$\begin{aligned} \boldsymbol{p} = \mu\boldsymbol{v} &= (\mu\dot{r})\,\hat{\boldsymbol{r}} + (\mu r\,\dot{\phi})\,\hat{\boldsymbol{\phi}} \\ &= \mu(\dot{r}\cos\phi - r\dot{\phi}\sin\phi)\hat{\boldsymbol{x}} + \mu(\dot{r}\sin\phi + r\dot{\phi}\cos\phi)\hat{\boldsymbol{y}}. \end{aligned} \tag{5.64}$$

The angular momentum is given by Equation (3.33):

$$\boldsymbol{L} = +(\mu r^2\dot{\phi})\hat{\boldsymbol{z}}. \tag{5.65}$$

Taking the cross product $\boldsymbol{p} \times \boldsymbol{L}$ gives

$$\begin{aligned} \boldsymbol{A} = &[\mu^2 r^2\dot{\phi}(\dot{r}\sin\phi + r\dot{\phi}\cos\phi) + \mu\kappa\cos\phi]\hat{\boldsymbol{x}} \\ &+ [-\mu^2 r^2\dot{\phi}(\dot{r}\cos\phi - r\dot{\phi}\sin\phi) + \mu\kappa\sin\phi]\hat{\boldsymbol{y}}. \end{aligned} \tag{5.66}$$

Now we need an expression for the radial velocity, $\dot{r}$. This was evaluated in Equation (5.12):

$$\begin{aligned} \frac{dr}{dt} &= -\frac{a(1-\varepsilon^2)\,\varepsilon\sin\phi}{(1-\varepsilon\cos\phi)^2}\left(\frac{d\phi}{dt}\right) = -\frac{r^2\varepsilon}{a(1-\varepsilon^2)}\sin\phi\left(\frac{d\phi}{dt}\right) \\ &= -\frac{\varepsilon L}{\mu a(1-\varepsilon^2)}\sin\phi, \end{aligned} \tag{5.67}$$

where the last step follows from Equation (3.35),

$$\dot{\phi} = \frac{L}{\mu r^2}. \tag{5.68}$$

Substitute Equations (5.67) and (5.68) into Equation (5.66). You will also need the equation for an ellipse, and set $\kappa = -GMm$. After some algebra, you should find that the components do indeed emerge as constants:

$$A_x = -\frac{\varepsilon L^2}{a(1-\varepsilon^2)}, \tag{5.69}$$

and

$$A_y = 0. \tag{5.70}$$

Alternate expressions for $A_x$ can be had by replacing $L^2$ with $L^2 = GMm\mu a(1 - \varepsilon^2)$:

$$A_x = -\varepsilon\mu GMm = -\frac{\varepsilon G(Mm)^2}{M + m} = +\kappa\varepsilon\mu. \tag{5.71}$$

So, $\boldsymbol{A}$ is constant and points purely in the negative $x$-direction. This shows that an ellipse must correspond to an inverse-square force.

There is a loose end to be addressed here. It was remarked just after Equation (5.19) that there are two conserved quantities in the physics of orbital motion, energy and angular momentum, and that two corresponding parameters, $a$ and $\varepsilon$, are needed to describe an ellipse. How then can *another* conserved quantity arise? The resolution of this question is that $A$ is not independent of $E$ and $L$. This can be seen by using Equation (5.32) for $E$ to eliminate $a$ in the expression for $L^2$ in Equation (5.10). Then eliminate $\varepsilon^2$ by utilizing Equation (5.71) immediately above in the form $A_x^2 = \kappa^2\varepsilon^2\mu^2$. The result is

$$A_x^2 = 2\mu EL^2 + \mu^2\kappa^2, \tag{5.72}$$

which shows that $A$ is fixed once $E$ and $L$ are known.

**Exercise:** What are the units of $\boldsymbol{A}$? Answer: $\mathrm{kg^2\,m^3\,s^{-2}}$.

## 5.8 Kepler's Third Law for Non-Inverse-Square Central Forces

In Section 3.7, it was shown that Newton's shell-point equivalency theorem also holds for a central force of the form $F(r) \propto r$, the harmonic oscillator force. In this section, Kepler's third law for central but non-inverse-square forces is considered. This is a hypothetical exercise, but reveals an additional interesting behavior in the case of the harmonic oscillator force.

For simplicity, the discussion here is restricted imagining circular orbits with $M \gg m$; we don't consider any motion of the central mass $M$. As in Section 3.7, imagine that the force is of the attractive form $\boldsymbol{F} = -Kr^{n-1}\hat{\boldsymbol{r}}$. The units of $K$ will depend on the value of $n$; the standard Newton/Kepler case corresponds to $K = GMm$ and $n = -1$.

If a planet of mass $m$ is in a circular orbit of speed $v$ and radius $r$ around a central "Sun," we know from basic dynamics that it must experience an attractive centripetal force of magnitude $mv^2/r$. This is presumed to be supplied by our revised gravitational force of *magnitude* $Kr^{n-1}$, so we can write

$$\frac{mv^2}{r} = Kr^{n-1} \Rightarrow v^2 = \frac{K}{m}r^n. \tag{5.73}$$

The period $T$ of the orbit must be related to $v$ and $r$ by $T = 2\pi r/v$. Kepler's third law expresses $T^2$ in terms of $r$, so in this case we have

$$T^2 = \frac{4\pi^2 r^2}{v^2} \Rightarrow T^2 = \frac{4\pi^2 m}{K} r^{2-n}. \tag{5.74}$$

Check that $K = GMm$ and $n = -1$ gives the conventional form of Kepler's Third Law for a circular orbit.

Now here is the interesting point. For the harmonic oscillator force, $n = 2$. In this case, *the orbital period is independent of the radius!* This is not as strange as it may sound: recall that the period of a simple harmonic oscillator is independent of its amplitude. This shows a connection (admittedly somewhat contrived) between basic mechanics and orbital motions.

It is also possible to show that a non-inverse-square central force cannot yield a closed orbit such as an ellipse; this is explored to some extent in Appendix B. A circular orbit such as posited here would consequently be unstable against any perturbation; such solar systems would likely not be suitable for any long-term evolution of life. There is nothing in fundamental physics that constrains gravity to be inverse-square, but if it were not, you would likely not be around to think about it!

**Exercise:** In a solar system in an alternate universe, one planet has an orbital radius of 2 AUs and a period of 3 years, while another has an orbital radius of 4 AUs and a period of 12 years. What is the value of $n$ for this system? Answer: $n = -2$, corresponding to an inverse-cube force.

## 5.9 The Effective Potential

The concept of effective potential was mentioned briefly in Section 3.6. In this section, this concept is used to illustrate as an alternate way of visualizing orbits.

See Equation (3.37). The effective potential $U(r)$ is defined as the sum of the potential energy and and the angular momentum term:

$$U(r) = V(r) + \frac{L^2}{2\mu r^2} = -\frac{GMm}{r} + \frac{L^2}{2\mu r^2}. \tag{5.75}$$

Strictly, this is an *effective potential energy*, but the custom is to call it just the "effective potential." With this definition, Equation (3.37) shows that the total energy $E$ of an orbital system can be written as the effective potential plus a term involving the square of the instantaneous radial velocity, $(dr/dt)$:

$$E = U(r) + \frac{\mu}{2}\left(\frac{dr}{dt}\right)^2. \tag{5.76}$$

Equation (5.75) is of the general form $U(r) = -g/r + h/r^2$ with $g = GMm$ and $h = L^2/2\mu$; note that $g$ is *not* the usual acceleration due to gravity. Any function of this general form has the shape shown as the solid curve sketched in Figure 5.3. When $r \to 0$, the inverse-square term $h/r^2$ dominates, so $U(r) \to +\infty$ in this regime. On the other hand, when $r \to \infty$, the $-g/r$ term does not go to zero as fast as the $h/r^2$

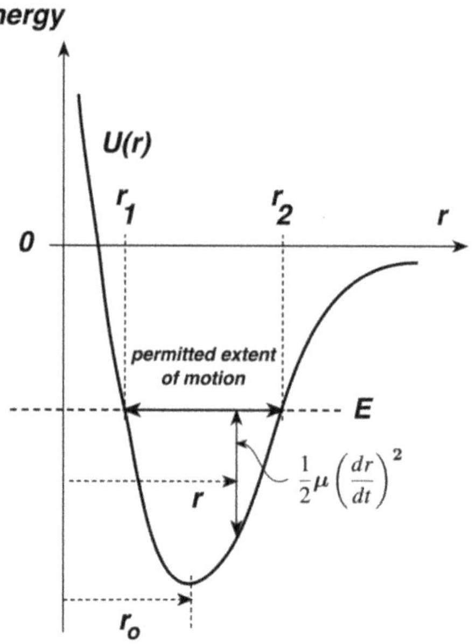


**Figure 5.3.** Sketch of the effective potential function $U(r)$. Not to scale. $r_o$ is the radial position where $U(r)$ is a minimum.

term does, so $U(r)$ approaches zero from below. $U(r)$ possesses one minimum, at $r = 2h/g = L^2/\mu GMm$.

Now suppose that the system has some total energy $E$; from Equation (5.19) we know that $E < 0$. This is drawn as a horizontal line in the figure. Note that $E$ cannot be more negative than the minimum value of $U(r)$, as this would imply a negative kinetic energy. At any general value of $r$, the vertical separation between the line for $E$ and the curve for $U(r)$ must be a measure of the radial velocity of the system from Equation (5.76): $E - U(r) = (\mu/2)(dr/dt)^2$.

The key point here is that the line for $E$ cuts the $U(r)$ curve at two radii; these are labeled $r_1$ and $r_2$ in the figure. Because we cannot have $E < U(r)$, these distances dictate the possible radial extent of the motion: apapsis and periapsis. This does not tell us the shape of the orbit, but it does tell us its distance limits. Conceptually, the curve for $U(r)$ is said to form a "potential well" within which the orbiting mass is confined. If you take a course in quantum mechanics, you are guaranteed to encounter the concept of a potential well.

At $r_1$ and $r_2$, $E = U(r)$, so the radial velocity $(dr/dt)$ must be zero at these positions. $r_1$ and $r_2$ are known as the "turning points" of the motion. If the orbiter has total energy equal to exactly the value of $U(r)$ at its minimum, it must always be at distance $r_o$, that is, it must be executing a circular orbit; the subscript "o" is used as a reminder of this. This orbit will be the concern of the next section; here we are interested in the more general energy and its corresponding limits $r_1$ and $r_2$.

The values of $r_1$ and $r_2$ can be found by seeking the intersection of the line for $E$ and the $U(r)$ curve:

$$E = -\frac{g}{r} + \frac{h}{r^2} \Rightarrow Er^2 + gr - h = 0. \tag{5.77}$$

This is a quadratic equation in $r$. The solution is

$$r = \frac{-g \pm \sqrt{g^2 + 4Eh}}{2E}. \tag{5.78}$$

On substituting for $g$ and $h$, setting $E = -GMm/2a$, and substituting $L^2 = GMm\mu a(1 - \varepsilon^2)$ from Equation (5.10) in the definition of $h$, you should find $r = a(1 \pm \varepsilon)$, exactly the conventional expressions for the periapsis and apapsis distances.

**Exercise:** The $r$-value of the bottom of the $U(r)$ curve must correspond to the position of greatest radial velocity. Show that this occurs at $r = a(1 - \varepsilon^2)$; you will need the above expression for $L^2$. Back-substitute this result into Equations (5.75) and (5.76) along with $E = -GMm/2a$ to show that

$$\left(\frac{dr}{dt}\right)_{\max} = \sqrt{\frac{G(M + m)\,\varepsilon^2}{a(1 - \varepsilon^2)}}.$$

## 5.10 A Taste of Perturbation Theory

Perturbation theory concerns what happens when an object in a stable orbit receives a sudden input of energy, a "perturbation." The general concept is that the amount of perturbing energy is small compared to the total energy of the orbit to begin with. The question is: what happens to the orbit? This section investigates this issue for the simple case of an initially circular orbit. This derivation builds on the analysis of effective potential developed in the preceding section. Our concern here is with the circular orbit of radius $r_0$ in Figure 5.3; we again ignore any motion of the presumed-large central mass, and use just an ordinary $m$ in place of the reduced mass $\mu$. In this analysis, we will stick to the $g$ and $h$-notation form of Equation (5.75); there will be no need to get into specific numerical values.

The effective potential is

$$U(r) = -\frac{g}{r} + \frac{h}{r^2}. \tag{5.79}$$

The minimum of $U(r)$ occurs at

$$r_0 = \frac{2h}{g}, \tag{5.80}$$

at which position the effective potential has the value

$$U(r_0) = -\frac{g^2}{4h}. \tag{5.81}$$

For a mass $m$ in a circular orbit of radius $r_o$, Kepler's third law gives the square of the period as

$$T_o^2 = \frac{32\pi^2 mh^3}{g^4}. \tag{5.82}$$

This comes from using Equation (5.80) for $r$ in $T^2 = 4\pi^2 r^3/GM$ and setting $GM = g/m$. We will return to this expression presently.

Now define a new radial coordinate $x$ as

$$x = r - r_o. \tag{5.83}$$

$x$ is *not* the usual Cartesian coordinate, it is used here to allow us to shift the origin of the radial coordinate so that $x = 0$ corresponds to the position of the minimum of the $U(r)$ curve. In terms of $x$,

$$U(x) = -\frac{g}{(x + r_o)} + \frac{h}{(x + r_o)^2}. \tag{5.84}$$

Now suppose that the orbiter receives a slight *purely radial* push inwards or outwards. We restrict ourselves to radial pushes because the definition of effective potential, Equations (5.75) and (5.76), concerns the radial motion $dr/dt$. The push will increase the orbiter's speed and hence its total energy, which means that it will go into a configuration akin to the dashed horizontal line in Figure 5.3. The premise here is to look at small pushes, that is, where the displacement $x$ is small compared to $r_o$, so $x/r_o \ll 1$ and the line for the total energy is displaced upwards only very slightly.

In each of the denominators in Equation (5.84), extract a factor of $r_o$:

$$U(x) = -\frac{g}{r_o(1 + x/r_o)} + \frac{h}{r_o^2(1 + x/r_o)^2}. \tag{5.85}$$

Perform binomial expansions:

$$U(x) \sim -\frac{g}{r_o}\left[1 - \frac{x}{r_o} + \left(\frac{x}{r_o}\right)^2 + \cdots\right] + \frac{h}{r_o^2}\left[1 - 2\left(\frac{x}{r_o}\right) + 3\left(\frac{x}{r_o}\right)^2 + \cdots\right]. \tag{5.86}$$

Gathering terms in the same power of $x$ gives

$$U(x) \sim \left[-\frac{g}{r_o} + \frac{h}{r_o^2}\right] + \left[\frac{g}{r_o^2} - 2\frac{h}{r_o^3}\right]x + \left[-\frac{g}{r_o^3} + 3\frac{h}{r_o^4}\right]x^2 + \cdots. \tag{5.87}$$

Substitute for $r_o$ from Equation (5.80). You will find the first term on the right side of (5.87) is just $U(r_o)$ of Equation (5.81). The middle term, that in the first power of $x$, vanishes entirely. This leaves (5.87) as

$$U(x) - U(r_o) \sim \left[\frac{g^4}{16h^3}\right]x^2 + \cdots. \tag{5.88}$$

For a reason that will become clear shortly, define a new constant $k$:

$$k = \left[\frac{g^4}{8h^3}\right]. \tag{5.89}$$

With this, Equation (5.88) becomes

$$U(x) \sim U(r_o) + \frac{1}{2}k\,x^2 + \cdots. \tag{5.90}$$

Now recall Equation (5.76) for the total energy of the system (replace $\mu$ with $m$):

$$E = U(r) + \frac{m}{2}\left(\frac{dr}{dt}\right)^2. \tag{5.91}$$

When we substitute (5.90) for $U$, we can cancel $E$ and $U(r_o)$ because $U(r_o)$ *is* the total energy for the unperturbed orbit, and the perturbation is presumed to be small. This leaves

$$0 = \frac{1}{2}kx^2 + \frac{m}{2}\left(\frac{dr}{dt}\right)^2. \tag{5.92}$$

Since $r = x + r_o$ and $r_o$ is a constant, $(dr/dt)^2 = (dx/dt)^2$, so we can write this as

$$m\left(\frac{dx}{dt}\right)^2 = -kx^2. \tag{5.93}$$

Take the time-derivative of this expression:

$$2\,m\left(\frac{dx}{dt}\right)\left(\frac{d^2x}{dt^2}\right) = -2kx\left(\frac{dx}{dt}\right). \tag{5.94}$$

Canceling the common factors of 2 and $(dx/dt)$ leaves, finally,

$$m\left(\frac{d^2x}{dt^2}\right) = -kx. \tag{5.95}$$

*Physical interpretation: This is exactly the form of the potential energy for a mass-spring system, $F = -kx = ma$. This tells us is that for slight radial perturbations of the circular orbit of radius $r_o$, the orbiting mass will exhibit simple harmonic motion in the radial direction about the "equilibrium position" $r_o$. This is the first main conclusion of our circular-orbit perturbation theory.*

Another quite remarkable conclusion can be drawn from this analysis concerning the period of the radial oscillations. From the mass-spring analogy, we know that the period for a full oscillation for a mass $m$ is

$$T_{osc} = 2\pi\sqrt{\frac{m}{k}}. \tag{5.96}$$

If you square this result and substitute for $k$ from Equation (5.89), you will find that to this level of approximation, *the oscillatory period is exactly equal to the period of the unperturbed orbit*, Equation (5.82). Since there was no perturbation to the *circular* motion, that is, to the angular motion, we can conclude that, again to this level of approximation, the orbit will remain closed, that is, it will always loop back on itself. Our circular orbit is thus stable against "small" radial perturbations.

As you might imagine, this analysis is but the tip of a very large iceberg; various forms of perturbation theory occur in various areas of physics. Appendix B expands on the development presented here to consider perturbations to circular orbits under the action of potentials of the more general form $V(r) = -kr^n$, where $n$ is presumed to be an integer; this was the type of general potential considered in the development of the shell-point equivalency theorem in Section 3.7. It is shown that for an orbit to be stable against radial perturbations (that is, to not have the orbiter run off to infinity), one must have $n > -2$, and also that, to have a closed orbit under the action of a perturbation, $\sqrt{n+2}$ must be equal to an integer. The case considered in this section, $n = -1$, satisfies this criterion.

## 5.11 Escape Velocity

Escape velocity is not an orbital issue per se, but can be dealt with very quickly via the potential energy function of Equation (3.23).

Consider a rocket of mass $m$ launched with initial speed $v_o$ from the surface of a planet of mass $M$ and radius $R$. After launch, the rocket's motor stops firing, and the rocket coasts thereafter. At the moment of launch, the total energy of the rocket/planet system will be the potential energy of the system plus the kinetic energy of the rocket:

$$E = -\frac{GMm}{R} + \frac{1}{2}mv_o^2. \tag{5.97}$$

Energy must be conserved, so at some later general distance $r$ from the center of the planet, the rocket will have some speed $v$ such that

$$-\frac{GMm}{R} + \frac{1}{2}mv_o^2 = -\frac{GMm}{r} + \frac{1}{2}mv^2. \tag{5.98}$$

The meaning of escape velocity is: what must be the initial velocity $v_o$ so that the rocket coasts out to $r = \infty$, arriving there just with $v = 0$? In this case, the right side of Equation (5.98) is zero, and the launch velocity is called the escape velocity:

$$v_{esc} = \sqrt{\frac{2GM}{R}}. \tag{5.99}$$

The escape velocity is independent of the mass of the rocket. Table 5.1 lists escape velocities for several solar-system objects. That for Earth is just over 11 kilometers per second, which is equivalent to almost exactly 25,000 miles per hour!

If the escape velocity should exceed the speed of light, then not even light would be able to escape: your gravitating object is then a black hole. For a mass $M$, the

**Table 5.1.** Escape Velocity Data.

| Object | Mass (Earth = 1) | Diameter (km) | Escape Velocity (km $s^{-1}$) |
|---|---|---|---|
| Sun | 332,946 | 1,391,400 | 617.7 |
| Mercury | 0.055 3 | 4879 | 4.3 |
| Venus | 0.815 0 | 12,104 | 10.4 |
| Earth | 1.000 0 | 12,756 | 11.2 |
| Mars | 0.107 5 | 6792 | 5.0 |
| Jupiter | 317.83 | 142,984 | 59.5 |
| Saturn | 95.159 | 120,536 | 35.5 |
| Uranus | 14.50 | 51,118 | 21.3 |
| Neptune | 17.20 | 49,528 | 23.5 |

$M_{\text{Earth}} = 5.972 \times 10^{24}$ kg.
Newtonian gravitational constant: $G = 6.674 \times 10^{-11}$ $m^3$ ($kg^{-1}$ $s^{-2}$).

radius to which the object needs to be compressed to make this so is known as the *Schwarzschild radius*:

$$R_{\text{Schwarz}} = \frac{2GM}{c^2}, \tag{5.100}$$

where $c$ is the speed of light. For Earth, $R_{\text{Schwarz}}$ is just under one centimeter.

**Exercise:** The diameter of the Moon is 3475 km, and its mass is 0.012 3 Earth masses. What is the escape velocity? Ans: 2.4 km $s^{-1}$.

AAS | IOP Astronomy

Keplerian Ellipses (Second Edition)
A student guide to the physics of the gravitational two-body problem
**Bruce Cameron Reed**

# Chapter 6

## Kepler's equation: Anomalies True, Eccentric, and Mean

With Equation (5.48) of the preceding chapter, we have an expression that relates elapsed time and the angular location $\phi$ of a planet in its orbit. Historically, much of astronomy was concerned with predicting when a planet would be at some position in the sky, usually for timekeeping and navigational purposes. To say that dealing with Equation (5.48) would have been a computational challenge in the days before even slide rules were available is an understatement. Kepler himself developed a much more compact alternate expression, which is now known as Kepler's equation. This derivation is predicated on constructing a sort of proxy circular orbit to stand in for the planet's true elliptical orbit, with an angle related to $\phi$ used to track position. This equation is derived and examined in this chapter.

See Figure 6.1, which shows a typical planetary elliptical orbit of semimajor axis $a$. Circumscribe the ellipse with a circle of radius $a$ centered on the geometric center of the ellipse. Then draw a vertical line perpendicular to the major axis that passes through the location of the planet and is extended until it intersects the circle. Finally, draw a line from the center of the circle to this intersection point; this must be of length $a$. Define $\psi$ as the angle that this new radial direction line makes with the major axis as measured at the center of the ellipse. This line does *not* pass through the planet. $\psi$ is known as the "eccentric anomaly" of the planet, and $\phi$ as the "true anomaly." In Greek mathematics, the word "anomaly" was synonymous with "angular position measured with respect to some reference direction," in this case the major axis. Note that when $\phi = 0$, $\psi = 0$ as well.

The approach here is to first get an expression relating $\phi$ and $\psi$. When written in terms of $\psi$, it turns out that the integral in Equation (5.46) can be solved analytically to yield a much simpler expression than Equation (5.48), although it is still transcendental and cannot be solved for $\psi(t)$.

doi:10.1088/2514-3433/acb430ch6  

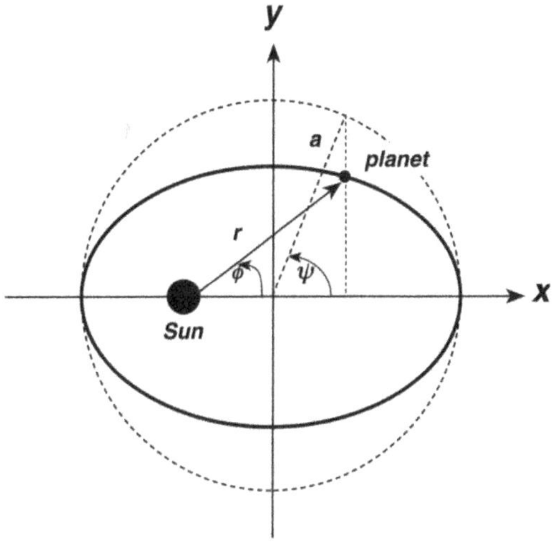


**Figure 6.1.** Construction for Kepler's equation.

Recall that the Sun is displaced from the center of the ellipse by distance $a\varepsilon$. From the $x$-coordinate of the planet, we can write

$$r \cos \varphi - a\varepsilon = a \cos \psi. \tag{6.1}$$

Substitute the expression for $r$ from the equation of an ellipse, (4.3), into this expression. The factors of $a$ cancel, leaving

$$\frac{(1 - \varepsilon^2) \cos \varphi}{1 - \varepsilon \cos \varphi} - \varepsilon = \cos \psi. \tag{6.2}$$

This can be rearranged to give

$$\cos \varphi = \frac{\varepsilon + \cos \psi}{1 + \varepsilon \cos \psi}. \tag{6.3}$$

Now, circular as this may seem (bad pun intended), back-substitute this expression into the equation of the ellipse. You should find that it reduces to

$$r = a(1 + \varepsilon \cos \psi). \tag{6.4}$$

If $\varepsilon = 0$, this is just the equation for a circle of radius $a$. This expression for the planet's distance from the Sun is certainly simpler than its "$\phi$" counterpart.

For a reason that will become apparent shortly, we will need an expression for the differential $d\phi$ in terms of $\varepsilon$, $\psi$, and $d\psi$. To get this, differentiate Equation (6.3):

$$-\sin \varphi \, d\varphi = \frac{-\sin \psi \, d\psi}{1 + \varepsilon \cos \psi} - \frac{(\varepsilon + \cos \psi)}{(1 + \varepsilon \cos \psi)^2}(-\varepsilon \sin \psi \, d\psi). \tag{6.5}$$

Bringing the right side of this to a common denominator gives

$$\sin \varphi \, d\varphi = \frac{(1 - \varepsilon^2) \sin \psi}{(1 + \varepsilon \cos \psi)^2} \, d\psi. \tag{6.6}$$

To isolate $d\phi$ now demands an expression for $\sin\phi$ in terms of $\varepsilon$ and $\psi$. This can be obtained by using Equation (6.3) and the identity $\sin^2\phi = 1 - \cos^2\phi$; the result is

$$\sin\varphi = \frac{\sqrt{1-\varepsilon^2}\sin\psi}{(1+\varepsilon\cos\psi)}. \tag{6.7}$$

With Equation (6.6), this gives

$$d\varphi = \frac{(1-\varepsilon^2)\sin\psi}{\sin\varphi(1+\varepsilon\cos\psi)^2}\,d\psi = \frac{\sqrt{1-\varepsilon^2}}{(1+\varepsilon\cos\psi)}d\psi. \tag{6.8}$$

We are now ready to get to Kepler's equation. This is done by setting up an integral for the time-angle equation equivalent to Equation (5.46), but by working directly with the radial distance $r$. Equation (5.46) arose from Equation (5.43), which is written here by putting $L/\mu$ in terms of the period and eccentricity; see the exercise at the end of Section 5.4 and use Kepler's third law to eliminate $(M+m)$:

$$\frac{d\varphi}{dt} = \frac{L}{\mu\, r^2} = \frac{2\pi a^2}{T\, r^2}\sqrt{1-\varepsilon^2}. \tag{6.9}$$

Separating variables and integrating gives

$$t_{0\to\varphi} = \frac{T}{2\pi\, a^2\sqrt{1-\varepsilon^2}}\int_0^{\varphi} r^2 d\varphi. \tag{6.10}$$

Now transform this integral to one over $\psi$ by using Equations (6.4) and (6.8); the result is

$$t_{0\to\psi} = \frac{T}{2\pi}\int_0^{\psi}(1+\varepsilon\cos\psi)d\psi. \tag{6.11}$$

This integral is trivial. It is customary in celestial mechanics research to define

$$\omega = \frac{2\pi}{T}, \tag{6.12}$$

in analogy to the angular frequency of a simple harmonic oscillator. With this, we can integrate Equation (6.11) to give

$$\omega\, t_{0\to\psi} = \psi + \varepsilon\sin\psi. \tag{6.13}$$

This is Kepler's equation; $\omega t$ is known as the mean anomaly. This result cannot be solved for $\psi(t)$, but it is considerably easier to work with than Equation (5.48). If $\varepsilon = 0$, this describes the angular position of an object moving at a uniform angular speed $\omega$ around a circle of radius $a$. $\psi$ must be expressed in radians.

For computational purposes, it can be handy to have a compact expression relating $\psi$ and $\phi$. This can be obtained from Equation (6.3) as follows. First, use that expression to set up expressions for $1 \pm \cos\phi$:

$$1 - \cos\phi = \frac{1 + \varepsilon\cos\psi - \varepsilon - \cos\psi}{1 + \varepsilon\cos\psi} = \frac{(1 - \varepsilon)(1 - \cos\psi)}{1 + \varepsilon\cos\psi} \tag{6.14}$$

and

$$1 + \cos\phi = \frac{1 + \varepsilon\cos\psi + \varepsilon + \cos\psi}{1 + \varepsilon\cos\psi} = \frac{(1 + \varepsilon)(1 + \cos\psi)}{1 + \varepsilon\cos\psi}. \tag{6.15}$$

The ratio of these gives

$$\frac{1 - \cos\phi}{1 + \cos\phi} = \frac{(1 - \varepsilon)(1 - \cos\psi)}{(1 + \varepsilon)(1 + \cos\psi)}. \tag{6.16}$$

The reason for this curious manipulation is that there is a trigonometric identity concerning the tangents of half-angles:

$$\tan\left(\frac{x}{2}\right) = \sqrt{\frac{1 - \cos x}{1 + \cos x}}. \tag{6.17}$$

Applying this to the square root of Equation (6.16) gives

$$\tan\left(\frac{\phi}{2}\right) = \sqrt{\frac{(1 - \varepsilon)}{(1 + \varepsilon)}}\tan\left(\frac{\psi}{2}\right). \tag{6.18}$$

When calculating inverse tangents, one normally has to be very careful to keep track of quadrants, but in this case we can be more cavalier. Most calculator and spreadsheet programs will return an angle corresponding to the so-called "principal value" of the argument of the inverse tangent. This means a result in either the first or fourth quadrants, depending on whether the argument is positive or negative. But there is always a second possible answer, with the other being the returned value plus 180°; it is up to a human being to decide which solution is relevant to the problem at hand. Now suppose that in evaluating Equation (6.18), the inverse tangent gives an angle of $\delta$ degrees. The other solution is then $\delta + 180°$. But this will be the result for $\phi/2$ or $\psi/2$, so $\phi$ or $\psi$ will be equal to either $2\delta$ or $2\delta + 360°$. But all trig functions are cyclical over 360°, so the two solutions correspond to the same direction! Sometimes the math does give you a break.

A comment to students on Equation (6.17). Most normal human beings do not memorize such identities. But do spend time flipping through tables of integrals and mathematical handbooks to get a sense of what identities are "out there." You never know when they might come in handy.

As an example of using Kepler's equation, we compute the time necessary for the spy satellite in the example at the end of the preceding chapter to cover $\phi = 0°$ to 90°. The eccentricity is $\varepsilon = 0.616$. $\phi = 0°$ corresponds to $t = 0$ and $\psi = 0°$; for $\phi = 90°$, $\cos\phi = 0$, and $\psi$ evaluates as

$$\psi = \cos^{-1}\left(\frac{\cos\varphi - \varepsilon}{1 - \varepsilon\cos\varphi}\right) = \cos^{-1}(-0.616) = 128°.\ 0 = 2.234\ \text{rad}.$$

Hence

$$\omega\, t_{\psi=128} = 2.234 + 0.616 \sin(2.234 \text{ rad}) = 2.719.$$

This gives a time of

$$t_{\psi=128} = \frac{2.719}{\omega} = \frac{2.719}{2\pi} T = 0.433\, T.$$

The travel time for $\phi = 270°$ back to 0° will be the same, for a total time $0.866\, T$ for $\phi = 270°$ to 90°. This leaves only $0.134\, T$ for the close-approach phase of $\phi = 90°$ to 270°, or about 48 minutes as claimed.

Along this line, Equations (6.4) and (6.13) provide a quick way to determine for what fraction of its orbit a planet will be within some heliocentric distance $r_{\text{helio}}$ of the Sun. Write the heliocentric distance in question as a fraction $\beta$ of the semimajor axis in the form $r_{\text{helio}} = \beta a$. The value of $\psi$ at which the object will reach this distance is given by Equation (6.4) as

$$\psi(r_{\text{helio}}) = \cos^{-1}\left(\frac{\beta - 1}{\varepsilon}\right). \tag{6.19}$$

To determine the time from perihelion at which this occurs, we need an expression for $\varepsilon \sin \psi$ for use in Equation (6.13). This can be obtained from

$$\varepsilon \sin \psi(r_{\text{helio}}) = \varepsilon\sqrt{1 - \cos^2 \psi} = \sqrt{\varepsilon^2 - (\beta - 1)^2}. \tag{6.20}$$

On accounting for both inbound and outbound travel, the fraction of the orbital period that the object will be at distances greater than $r_{\text{helio}}$ will be, from Equations (6.12) and (6.13),

$$\left(\frac{t}{T}\right)_{>r(\text{helio})} = \frac{1}{\pi}\left[\cos^{-1}\left(\frac{\beta - 1}{\varepsilon}\right) + \sqrt{\varepsilon^2 - (\beta - 1)^2}\right]. \tag{6.21}$$

The fraction of the period spent within $r_{\text{helio}}$ will thus be

$$\left(\frac{t}{T}\right)_{<r(\text{helio})} = 1 - \frac{1}{\pi}\left[\cos^{-1}\left(\frac{\beta - 1}{\varepsilon}\right) + \sqrt{\varepsilon^2 - (\beta - 1)^2}\right]. \tag{6.22}$$

Again, do all calculations in radians. As an example, Halley's comet has $a = 17.834$ AU and $\varepsilon = 0.967\,1$. Of its 75.32-year period, how many days does it spend within $r_{\text{helio}} = 1$ AU of the Sun, that is, within the orbit of the Earth? These numbers give $\beta = 1/17.834 = 5.607 \times 10^{-2}$. You should find $(t/T)_{<r(\text{helio})} = 2.837 \times 10^{-3}$, corresponding to a mere 78 days! This example is somewhat contrived in that the plane of Halley's orbit is inclined at about 18° to that of Earth's, but does demonstrate the effect of the extreme eccentricity.

**Exercise:** Pluto's orbit has $a = 39.482$ AU and $\varepsilon = 0.248\,8$. What fraction of its orbit is spent within the semimajor axis of Neptune's orbit, $a = 30.07$ AU?

Answer: 0.069 75, or about 17.3 years of its period of 248 years. They will not collide, however, as Pluto's orbit is inclined by about 17° to that of Neptune's.

**Exercise:** Show that Equation (5.48) is equivalent to Equation (6.13). Equations (6.7) and (6.18) may be helpful.

# Chapter 7

## Transfer and Rendezvous Orbits

In the second half of the twentieth century, orbital mechanics experienced a revival of interest with the launch of satellites, probes to the Moon and planets, and, eventually, manned space flights. In these situations, the focus of analyses was often that of a "two-body" problem wherein it would be desired to arrange for a satellite to arrive at a prescribed place at some time. Examples could include sending a probe to Mars, intercepting an adversary satellite, or arranging for a rendezvous between two craft such as a supply vessel and a space station.

This chapter examines three such transfer-and-rendezvous problems. The first, taken up in Section 7.1, is known as the Hohmann ellipse transfer orbit. This maneuver was described in 1925 by German engineer Walter Hohmann (1880–1945), and involves a fuel-efficient elliptical path between two bodies orbiting a central mass, such as two planets orbiting the Sun. The specific problem is to compute necessary departure and arrival velocity changes to first send the probe into the proper trajectory and then to adjust its speed upon arrival to match the orbital speed of the target object. This procedure is advantageous in that it can minimize fuel use, but the disadvantage is that one has no control over the transfer time. In contrast, the Lambert transfer orbit (Section 7.2), named after Swiss/French mathematician Johann Lambert (1728–1777), can be very expensive in fuel use but allows a mission planner to specify the transit time. This can be useful if a mission is time-sensitive, such as having to intercept an attacking object. The third problem (Section 7.3), the "Ham sandwich transfer," sounds more whimsical but is quite serious. Imagine two astronauts orbiting the Earth in the same circular orbit but riding in separate capsules. One is somewhat ahead of the other, perhaps by as much as a couple hundred kilometers. How does the trailing astronaut throw his or her colleague a sandwich? The answer proves to be counterintuitive, involving a backwards throw!

doi:10.1088/2514-3433/acb430ch7

## 7.1 The Hohmann Ellipse Transfer Orbit

Suppose that you wish to send a probe to an inner (inferior) planet, say Venus. The premise of the Hohmann calculation is that the probe is first put into a circular orbit around the Sun of radius equal to that of Earth's orbit, $R_E$. (Note that $R_E$ was previously used to represent the radius of the Earth itself; here it is Earth's orbital radius.) It is important to emphasize that the maneuvers and fuel cost to achieve this initial circumstance is *not* considered here; these are taken to be separate issues. The idea is that the probe is arranged to be well away from Earth and its gravitational influence; the Sun plays the role of the "gravitating" mass.

This analysis will involve the circular speeds of both planets in their orbits around the Sun. For a planet of orbital radius $R$, this is given by

$$v_{\text{circ}} = \sqrt{\frac{GM_{\text{Sun}}}{R}}. \tag{7.1}$$

Let the inner planet have orbital radius $R_{\text{inner}}$. Consult Figure 7.1. At the moment when the probe is at the right side of the figure, an onboard rocket is fired to alter the probe's speed so that it is equal to that of the speed at aphelion of an elliptical orbit of semimajor axis $a = (R_E + R_{\text{inner}})/2$. This means that the aphelion distance of the transfer ellipse will be $R_E$, and, when it reaches perihelion, it will be at distance $R_{\text{inner}}$ from the Sun as shown. When it arrives at this point the rocket is fired again to bring its speed to $v_{\text{circ}}^{\text{inner}}$ in preparation for a further burn to drop it into an orbit around the target planet, or whatever is desired.

The Hohmann problem is to compute the necessary departure and arrival velocity alterations.

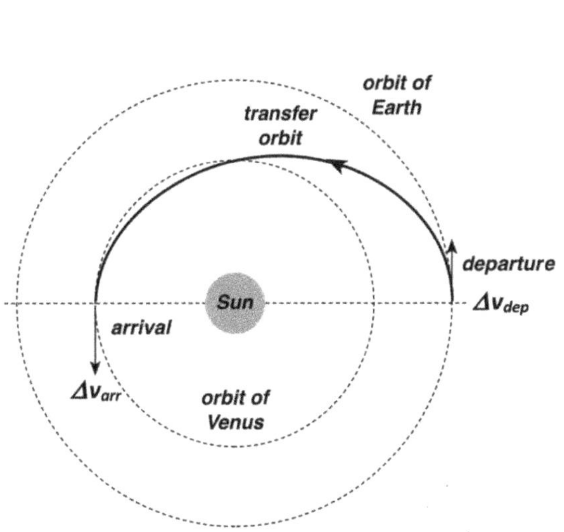


**Figure 7.1.** Hohmann ellipse transfer orbit from Earth to Venus. Not to scale.

What will be very useful here are Equations (5.16) and (5.17) for the aphelion and perihelion speeds of an elliptical orbit. These are written here assuming that the mass of the Sun is much greater than that of the probe:

$$v_{\text{aphelion}} = \sqrt{\frac{GM_{\text{Sun}}}{a}\frac{(1-\varepsilon)}{(1+\varepsilon)}} \tag{7.2}$$

and

$$v_{\text{perihelion}} = \sqrt{\frac{GM_{\text{Sun}}}{a}\frac{(1+\varepsilon)}{(1-\varepsilon)}}. \tag{7.3}$$

Since the aphelion and perihelion distances $a(1+\varepsilon)$ and $a(1-\varepsilon)$ are, respectively, $R_E$ and $R_{\text{inner}}$, we can set

$$\frac{(1+\varepsilon)}{(1-\varepsilon)} = \frac{R_E}{R_{\text{inner}}}. \tag{7.4}$$

With $a = (R_E + R_{\text{inner}})/2$, we can write Equations (7.2) and (7.3) as

$$v_{\text{aphelion}} = \sqrt{\frac{2GM_{\text{Sun}}}{R_E + R_{\text{inner}}}\left(\frac{R_{\text{inner}}}{R_E}\right)} \tag{7.5}$$

and

$$v_{\text{perihelion}} = \sqrt{\frac{2GM_{\text{Sun}}}{R_E + R_{\text{inner}}}\left(\frac{R_E}{R_{\text{inner}}}\right)}. \tag{7.6}$$

Now consider the velocity alteration at the departure (aphelion) point. The initial speed will be $v_{\text{circ}}^{\text{Earth}}$, but we want it to be $v_{\text{aphelion}}$. The change must be

$$\begin{aligned}\Delta v_{\text{depart}} = v_{\text{aphelion}} - v_{\text{circ}}^{\text{Earth}} &= \sqrt{\frac{2GM_{\text{Sun}}}{R_E + R_{\text{inner}}}\left(\frac{R_{\text{inner}}}{R_E}\right)} - \sqrt{\frac{GM_{\text{Sun}}}{R_E}} \\ &= \sqrt{\frac{GM_{\text{Sun}}}{R_E}}\left[\sqrt{\frac{2R_{\text{inner}}}{R_E + R_{\text{inner}}}} - 1\right] \\ &= v_{\text{circ}}^{\text{Earth}}\left[\sqrt{\frac{2}{x+1}} - 1\right],\end{aligned} \tag{7.7}$$

where

$$x = \frac{R_E}{R_{\text{inner}}}. \tag{7.8}$$

We must have $x > 1$, so $\Delta v_{\text{depart}}$ must be negative: you will need a retro-rocket. A convenience of this scheme is that all velocity alterations are either parallel or antiparallel to the instantaneous direction of travel.

At arrival, the issue is to change from speed $v_{\text{perihelion}}$ to $v_{\text{circ}}^{\text{inner}}$. The algebra is then reversed from the above:

$$\begin{aligned}\Delta v_{\text{arrive}} = v_{\text{circ}}^{\text{inner}} - v_{\text{perihelion}} &= \sqrt{\frac{GM_{\text{Sun}}}{R_{\text{inner}}}} - \sqrt{\frac{2GM_{\text{Sun}}}{R_{\text{E}} + R_{\text{inner}}}\left(\frac{R_{\text{E}}}{R_{\text{inner}}}\right)} \\ &= \sqrt{\frac{GM_{\text{Sun}}}{R_{\text{inner}}}}\left[1 - \sqrt{\frac{2R_{\text{E}}}{R_{\text{E}} + R_{\text{inner}}}}\right] \\ &= v_{\text{circ}}^{\text{inner}}\left[1 - \sqrt{\frac{2x}{x+1}}\right],\end{aligned} \tag{7.9}$$

where $x$ is defined as in Equation (7.8). Like $\Delta v_{\text{depart}}$, $\Delta v_{\text{arrive}}$ is also negative. Note that the prefactor in Equation (7.7) is $v_{\text{circ}}^{\text{Earth}}$, while that in Equation (7.9) is $v_{\text{circ}}^{\text{inner}}$.

The transit time will be exactly one-half of the period of an elliptical orbit of semimajor axis $a = (R_{\text{E}} + R_{\text{inner}})/2$.

In the case of a transfer to an outer (superior) planet, the situation is similar except that Earth's orbit now represents the perihelion distance and the target planet's the aphelion distance. Here the velocity alterations are

$$\Delta v_{\text{depart}} = v_{\text{perihelion}} - v_{\text{circ}}^{\text{Earth}} = v_{\text{circ}}^{\text{Earth}}\left[\sqrt{\frac{2y}{1+y}} - 1\right] \tag{7.10}$$

and

$$\Delta v_{\text{arrive}} = v_{\text{circ}}^{\text{outer}} - v_{\text{aphelion}} = v_{\text{circ}}^{\text{outer}}\left[1 - \sqrt{\frac{2}{1+y}}\right], \tag{7.11}$$

where

$$y = \frac{R_{\text{outer}}}{R_{\text{E}}}. \tag{7.12}$$

In this case, both velocity adjustments are positive: Speed boosts are required for outer planets.

Table 7.1 lists circular velocities and velocity adjustments for Earth-to-planet scenarios; note that $GM_{\text{Sun}} = 1.327 \times 10^{20}\ \text{m}^3\ \text{s}^{-2}$. Pluto is not considered here on account of its high eccentricity.

While Hohmann ellipses involve fairly modest velocity changes, they are impractical for missions to outer planets given the large transit times. In these cases, extra propulsion or gravity assists by arranging to swing past an intervening planet are utilized; see Section 8.9. The value of the Hohmann calculation is largely pedagogical.

**Table 7.1.** Hohmann Ellipse Transfer Data: Earth Orbit to Target Planet.

| Target planet | $R$ (AU) | $v_{circ}$ (km s$^{-1}$) | $\Delta v_{depart}$ (km s$^{-1}$) | $\Delta v_{arrive}$ (km s$^{-1}$) | Transit time |
|---|---|---|---|---|---|
| Mercury | 0.387 1 | 47.86 | −7.53 | −9.61 | 105 days |
| Venus | 0.723 3 | 35.02 | −2.50 | −2.71 | 146 days |
| Mars | 1.523 7 | 24.13 | 2.94 | 2.65 | 0.71 yr |
| Jupiter | 5.202 9 | 13.06 | 8.79 | 5.64 | 2.73 yr |
| Saturn | 9.537 | 9.64 | 10.29 | 5.44 | 6.05 yr |
| Uranus | 19.189 | 6.80 | 11.28 | 4.66 | 16.04 yr |
| Neptune | 30.07 | 5.43 | 11.65 | 4.05 | 30.62 yr |

**Exercise:** Show that $\Delta v_{arrive}$ from the Earth to an outer planet is a maximum for a planet whose orbital radius is about 5.9 AU, and has a value of about 5.66 km s$^{-1}$. You will have to solve numerically. This distance is just somewhat beyond the orbital radius of Jupiter ($R \sim 5.2$ AU).

**Exercise:** Inhabitants of Jupiter dispatch a Hohmann-ellipse probe to Uranus. What are the velocity changes and transit time? Answer: $\Delta v_{depart} \sim 3.32$ km s$^{-1}$; $\Delta v_{arrive} \sim 2.36$ km s$^{-1}$; ~21.3 years. Why is this latter result so large?

## 7.2 The Lambert Problem

In contrast to the Hohmann problem described in the preceding section, the Lambert trajectory allows a mission planner some control over the transfer time. While conceived in the 1760s, this problem is very relevant in current spaceflight dynamics when it is desired to give a craft a trajectory that gets it to a certain place in a certain time. For example, a supply rocket or probe may be required to rendezvous with a space station or planet at some particular time in the future, or one may have a limited amount of time to intercept an asteroid headed for a collision with Earth.

The analytic and numerical literature on the Lambert problem is extensive; see https://en.wikipedia.org/wiki/Lambert%27s_problem. However, analytic solutions demand considerable knowledge of conic sections, and numerical solutions are inherently iterative and often involve various auxiliary quantities. In view of these complications, it is probably not surprising that the Lambert problem is not a staple of textbook exercises. In this section I describe a spreadsheet-based approach to the Lambert problem that is very easy for even novice users to implement.

The Lambert problem can be described as follows. See Figure 7.2. There are three objects involved: A central attractor of mass $M$ (often the Earth or the Sun), a craft which can be directed (which I will refer to as the "probe"), and a target object. Suppose that it is known that at some moment the probe is at distance $r_1$ from the attractor. It is also known that at time $\tau$ later, the target will be at distance $r_2$ from the attractor. Since the directions of these positions are known, the angle $\gamma$ between the vectors $\boldsymbol{r}_1$ and $\boldsymbol{r}_2$ is known. However, at $\boldsymbol{r}_1$ the probe is *not* in an orbit that will make it intercept the target after time $\tau$. The question is: what are the parameters of

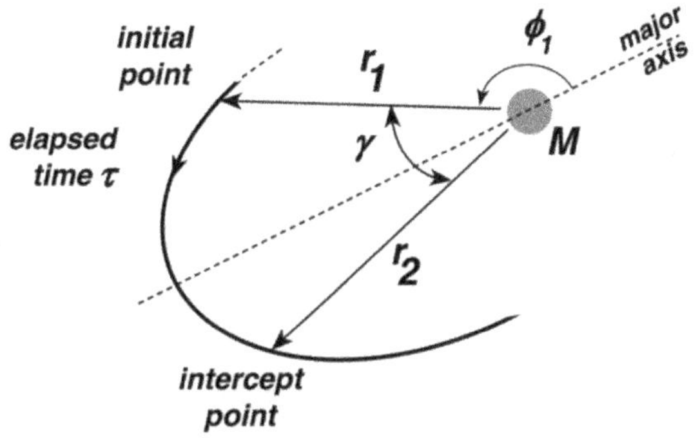


**Figure 7.2.** Geometry of the Lambert problem. Known quantities are the initial and rendezvous distances $r_1$ and $r_2$, the elapsed apsidal angle $\gamma$, the time of flight $\tau$, and the mass $M$ of the attractor.

the elliptical orbit about the attractor into which the probe needs to be put to make the rendezvous? Specifically, we wish to know the semimajor axis $a$, eccentricity $\varepsilon$, velocity $\boldsymbol{v}$ with respect to the attractor which the probe has to be given at $\boldsymbol{r}_1$, and the apsidal angle $\phi_1$ of the position $\boldsymbol{r}_1$ of the required orbit. The apsidal angle at $\boldsymbol{r}_2$ will be $\phi_2 = \phi_1 + \gamma$, so it will be known once $\phi_1$ has been established. We assume that the mass of the attractor is much greater than that of the probe, so that $G(M + m)$ can always be approximated as $GM$.

It is important to emphasize that the probe is *not* in the requisite orbit to begin with. Otherwise, no velocity adjustment would be necessary; this is why the trajectory of the probe is not drawn as a complete ellipse. The incoming trajectory of the target—which is not sketched at all—is a completely separate issue; all that is known of the target is that it will arrive at $\boldsymbol{r}_2$ a time $\tau$ seconds after the probe is at $\boldsymbol{r}_1$.

The approach taken here is to formulate the information on distances and travel time as an equation of constraint expressed in terms of the initial apsidal angle $\phi_1$, and to plot this equation as a function of $\phi_1$ over a range of values of $\phi_1$ in order to graphically determine the value of $\phi_1$ which satisfies the conditions of the problem.

Now, we know the distances $r_1$ and $r_2$. These must have corresponding angular positions $\phi_1$ and $\phi_2$. If the eccentricity of the orbit is $\varepsilon$ as usual, we can write

$$r_1 = \frac{a(1 - \varepsilon^2)}{1 - \varepsilon \cos \phi_1}, \tag{7.13}$$

and

$$r_2 = \frac{a(1 - \varepsilon^2)}{1 - \varepsilon \cos \phi_2}. \tag{7.14}$$

Now define $\rho$ as the ratio

$$\rho = \frac{r_1}{r_2} = \frac{1 - \varepsilon \cos \phi_2}{1 - \varepsilon \cos \phi_1}. \tag{7.15}$$

If a trial value of $\phi_1$ is chosen, then $\phi_2 = \phi_1 + \gamma$, and we can solve for $\varepsilon$:

$$\varepsilon = \frac{1 - \rho}{\cos \phi_2 - \rho \cos \phi_1}. \tag{7.16}$$

To formulate a similar constraint for the travel time, it is convenient to use *Kepler*'s equation, (6.13), which gives the time for the probe to travel from apapsis when the eccentric anomaly angle $\psi$ is zero to some general value of $\psi$ as

$$t = \frac{T}{2\pi}(\psi + \epsilon \sin \psi) = \frac{a^{3/2}}{\sqrt{GM}}(\psi + \epsilon \sin \psi), \tag{7.17}$$

where $T$ is the period of the orbit, $T^2 = 4\pi^2 a^3/GM$.

It will be necessary to relate $\psi$ and $\phi$, which Equation (6.18) gives as

$$\tan\left(\frac{\phi}{2}\right) = \sqrt{\frac{(1 - \varepsilon)}{(1 + \varepsilon)}} \tan\left(\frac{\psi}{2}\right). \tag{7.18}$$

Writing Equation (7.17) for positions 1 and 2 and taking the difference $t_2 - t_1 = \tau$ gives

$$\frac{a^{3/2}}{\sqrt{GM}}[(\psi_2 - \psi_1) + \epsilon(\sin \psi_2 - \sin \psi_1)] - \tau = 0. \tag{7.19}$$

For numerical purposes, it is more convenient to deal with quantities of order unity. To make Equation (7.19) convenient in this way, divide through by the mission time $\tau$. I define the resulting expression as $F(t)$ as it expresses the time constraint:

$$F(t) = \frac{a^{3/2}}{\tau \sqrt{GM}}[(\psi_2 - \psi_1) + \epsilon(\sin \psi_2 - \sin \psi_1)] - 1 = 0. \tag{7.20}$$

Fundamentally, Equation (7.20) is an equation of constraint in $\phi_1$: Once a value of $\phi_1$ is chosen, $\phi_2$ is known, $\varepsilon$ is known from Equation (7.16), and $\psi_1$ and $\psi_2$ are known from Equation (7.18). The easiest way to pin down the relevant value $\phi_1$ is to plot $F(t)$ versus $\phi_1$ and locate the value of $\phi_1$ which gives $F(t) = 0$.

In problems of this nature, it is helpful to use a convenient system of units. For example, if the Sun is the attractor, it is more convenient to have distances in Astronomical Units (AUs) and times in days. For any object, $GM$ has units of $(\text{distance})^3/(\text{time})^2$, and can be converted to the convenient units. For the Sun (check these numbers!)

$$GM_{\odot} = 1.327\,124 \times 10^{20} \text{m}^3\,\text{s}^{-2} = 2.959\,122 \times 10^{-4} \text{AU}^3\,\text{day}^{-2}.$$

Here is an example using these units. A probe initially at $r_1 = 1$ AU from the Sun is to be directed to rendezvous with Jupiter at $r_2 = 5.203$ AU after a flight of 1500 days (~4.1 years) that covers apsidal angle $\gamma = 150°$.

In any calculation of this nature, you will find that some values of $\phi_1$ will lead to values of the eccentricity which are either $<0$ (unphysical) or $>1$ (unbound orbit).

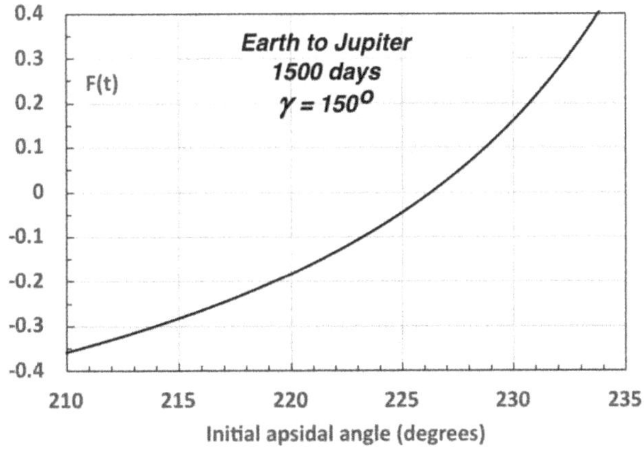


**Figure 7.3.** $F(t)$ of Equation (7.20) for an Earth-to-Jupiter mission with the Sun as the attractor body: $r_1 = 1$ AU, $r_2 = 5.203$ AU, $\tau = 1500$ days, and $\gamma = 150°$.

In the present case, you should find that only the range $\sim 160° \leqslant \phi_1 \leqslant \sim 250°$ gives values of $\varepsilon$ between 0 and 1. Figure 7.3 shows the run of $F(t)$ for $\sim 210° \leqslant \phi_1 \leqslant \sim 235°$; $F(t) = 0$ at $\phi_1 \sim 226°. 35$, which corresponds to $\varepsilon = 0.739$. The semimajor axis is about 3.33 AU.

The velocity vector can be formed from the radial and tangential components evaluated in Equations (5.13) and (5.14), taking $M \gg m$:

$$v = \dot{r}\,\hat{r} + (r\dot{\phi})\hat{\phi} = \sqrt{\frac{GM}{a(1 - \varepsilon^2)}}\,[(-\epsilon \sin\phi)\,\hat{r} + (1 - \epsilon\cos\phi)\hat{\phi}]. \tag{7.21}$$

For this example, the components evaluate to about 13.0 and 36.6 km s$^{-1}$ at $\phi_1$.

One should also compute the energy per unit mass of the probe, $E/m = v_1^2/2 - GM/r_1$, to verify that the orbit is not unbound; also, if the orbit should be in the vicinity of the attractor, one may have to check that a crash does not occur!

**Exercise:** A defense satellite in orbit around the Earth at an altitude of one Earth radius ($R_E = 6370$ km) above the surface detects an incoming asteroid. The satellite can intercept the asteroid at a distance of five Earth radii from the center of the Earth with a flight of $\tau = 4$ hours over an apsidal angle of $\gamma = 120°$. What are the necessary $\phi_1$ (to the nearest degree), $\varepsilon$, $a$, and velocity components? Take $GM = 398{,}600$ km$^3$ s$^{-2}$. Answer: $\phi_1 \sim 244°$, $\varepsilon \sim 0.512$, $a \sim 21{,}125$ km, $v_1 = (2.32\,\hat{r} + 6.19\,\hat{\phi})$ km s$^{-1}$. The orbit is bound.

## 7.3 The Ham Sandwich Throw

This section takes up an orbital transfer question posed many years ago but about which little seems to have been published: How can an astronaut in a circular orbit around the Earth toss a sandwich to a friend in another capsule in the same orbit but who is flying ahead of or behind them?

This problem was originally posed by California Institute of Technology physicist Dr. Lee DuBridge in an after-dinner speech at an American Physical Society meeting in 1960. A transcript of the talk was published in *American Journal of Physics* late that year, and is still well worth reading; his analysis of the unexpected results which occur if the thrower attempts a forward throw is a gem of humor and insight (DuBridge 1960). An excerpt:

*Imagine two spacecraft buzzing along in the same circular orbit around the Earth—say 400 miles up—and one ship is 100 yards or so ahead of the other one. The fellow in the rear vehicle wants to throw a baseball or a monkey wrench or a ham sandwich, or something, to the fellow ahead of him. How does he do it?*

*It sounds real easy. Since the two ships are in the same orbit, they must be going the same speed—so the man in the rear could give the baseball a good throw forward and the fellow ahead should catch it.*

*But wait! When you throw the ball out, its speed is added to the speed of the vehicle so now it is going too fast for that orbit. The centrifugal force (sic) is too great and the ball goes off on a tangent and rises to a higher orbit. But an object in a higher orbit must go slower. In fact, the faster he throws the ball, the higher it rises and the slower it goes. So our baseball pitcher stares in bewilderment as the ball rises ahead of him, then seems to stop, go back over his head, and recede slowly but surely to the rear, captured forever in a higher and slower and more elliptical orbit ⋯.*

Most of DuBridge's talk concerned more serious aspects of space exploration, but the underlying message was serious: within a few years astronauts would be dealing with real-life rendezvous issues.

This question has subsequently appeared as an end-of chapter problem in a few texts, but is usually posed as a qualitative exercise. A lecture on this issue by Massachusetts Institute of Technology astrophysicist Dr. Walter Lewin can be found online; his approach is the basis of the more general analysis offered here. Lewin's talk can be found at https://www.youtube.com/watch?v=Ky2XIElijvs.

The reason for a dearth of literature on this question is likely that one has to be careful with maintaining numerical precision because the speeds involved have quite different orders of magnitude. A 1960-era slide-rule and table-of-logarithms calculation would have been awkward, but an analysis is now straightforward thanks to the precision available with spreadsheets or even a hand calculator.

Expressed in more current parlance, the ham sandwich problem can be described as follows. See Figure 7.4. Alice (A) and Bob (B) are both in circular orbits of radius $R_o$ and speed $v_o$ about the Earth. Alice has a lead of $\theta_o$ radians (hence distance $R_o\theta_o$) on Bob; the case where she lags him is discussed below. He has prepared lunch and wishes to throw her a sandwich. How can he do so? Without any loss of generality, we can imagine the throw to occur when Bob crosses the $x$-axis as shown. Designate their common orbital period as $T_o$.

An obvious solution is for Bob to throw the sandwich with speed $2v_o$ *backwards relative to himself* so that it enters a reverse circular orbit of the same radius. In this case Alice will receive the sandwich after she has traveled less than half an orbit. However, this is not a practical solution. For an orbit of altitude 200 km, $v_o \sim 7800 \text{ m s}^{-1}$.

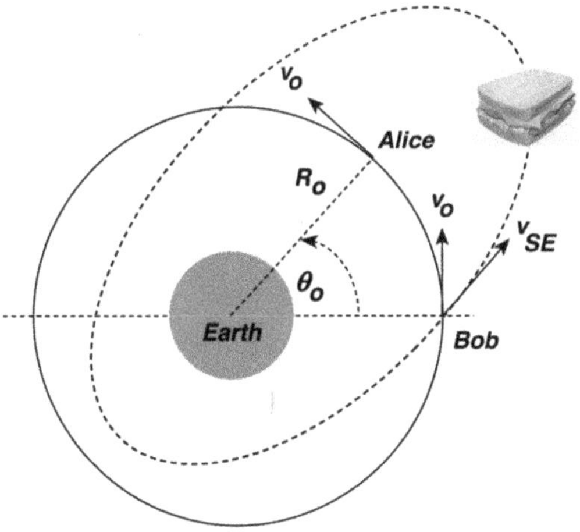


**Figure 7.4.** Ham sandwich transfer ellipse. The circular orbital velocity of both A and B is $v_o$, and the velocity of the sandwich relative to Earth at the moment of the throw is $v_{SE}$. Alice's lead, their common altitude, and the eccentricity of the ellipse are greatly exaggerated in this sketch.

A professional baseball pitcher can achieve a throw of ~40 m s$^{-1}$, only about 0.5% as much, so Bob will need a more effective strategy.

When Bob throws the sandwich, it will enter its own orbit around the Earth, and after one orbital period it will return to the throw point. If we arrange for the period to be the time it will take Alice to circle around to that point, she will intercept it. Her orbital speed is $2\pi/T_o$ radians per second, and to get around to the throw point she has to travel through $2\pi - \theta_o$ radians. Her travel time will then be $(2\pi - \theta_o)/(2\pi/T_o) = T_o(1 - \theta_o/2\pi)$. This is the necessary period of the sandwich's elliptical orbit:

$$T_{ell} = T_o(1 - \theta_o/2\pi). \tag{7.22}$$

One could arrange for Alice to catch the sandwich after executing multiple orbits, but I will not get into this here. The essential question is: if Bob can throw the sandwich with speed $v_{SB}$ relative to himself, in what direction $\theta_{SB}$ relative to himself should he make the toss to get the correct orbit? $\theta_{SB}$ is measured counterclockwise from $x$ as usual. If we write *Kepler*'s third law for both the initial circular orbit and the new elliptical one, $T_o^2 = 4\pi^2 R_o^3/GM$ and $T_{ell}^2 = 4\pi^2 a^3/GM$, where $M$ is the mass of the Earth and $a$ the semimajor axis of the sandwich orbit, we can eliminate $GM$ to get $a$ in terms of $R_o$:

$$a^3 = R_o^3 (1 - \theta_o/2\pi)^2. \tag{7.23}$$

The speed of the sandwich relative to Earth at the moment of launch, $v_{SE}$, then follows from the vis-viva equation ($M \gg m$):

$$v_{SE}^2 = GM\left(\frac{2}{R_o} - \frac{1}{a}\right). \tag{7.24}$$

Now, the launch velocity of the sandwich relative to Earth will be, from the usual procedure for combining vector velocities,

$$\boldsymbol{v}_{SE} = \boldsymbol{v}_{SB} + \boldsymbol{v}_{BE} = (v_{SB} \cos\theta_{SB})\hat{\boldsymbol{x}} + (v_{SB} \sin\theta_{SB} + v_o)\hat{\boldsymbol{y}}. \tag{7.25}$$

Squaring and summing the components allows us to extract the throw angle as

$$\sin\theta_{SB} = \frac{v_{SE}^2 - v_o^2 - v_{SB}^2}{2\, v_o\, v_{SB}}. \tag{7.26}$$

Before getting into numbers, some comments of general nature are worthwhile. First, for any practical scenario we will have $v_{SB} \ll v_o$, so even if Bob throws the sandwich backwards relative to himself it will appear to an outside observer to still be moving forward but perhaps with only a slight sideways component to its motion. It speed will be but little changed from $v_o$, so the semimajor axis will be only slightly different from $R_o$; the eccentricity of the orbit will not be nearly what is implied in Figure 7.4, which is drawn for clarity. Also, Equation (7.23) tells us that $a < R_o$, which in Equation (7.24) tells us that we must have $v_{SE} < v_o$, since (recall that $v_o^2 = GM/R_o$):

$$v_{SE}^2 - v_o^2 = GM\left(\frac{2}{R_o} - \frac{1}{a}\right) - \frac{GM}{R_o} = GM\left(\frac{1}{R_o} - \frac{1}{a}\right). \tag{7.27}$$

This means that the numerator of Equation (7.26) will always be negative (remember that $a < R_o$): Bob *must* throw the sandwich backwards relative to himself, although he can throw into the third or fourth quadrant as he pleases; his direction of forward motion at the moment of the throw is that of the $y$-axis. I will deal here with outwardly-directed (fourth quadrant) throws; if the throw were inward, the bottom of the ellipse in Figure 7.4 would point to the lower right instead of the lower left.

Figure 7.5 shows calculated throw angles as a function of Alice's lead distance for various throwing speeds for $M = 5.972 \times 10^{24}$ kg, $G = 6.674 \times 10^{-11}\ \mathrm{m^3\ kg^{-1}\ s^{-2}}$, and an altitude of 200 km above an Earth of radius 6370 km. A leisurely toss will reach Alice even if she is tens of kilometers ahead, although she will have to wait almost a full orbit before enjoying lunch. Bob will have to be very accurate with his aim and throw speed.

A situation to bear in mind here is that if the eccentricity $\varepsilon$ of the sandwich orbit is too large, it will crash into the Earth if its perigee distance $a(1 - \varepsilon)$ falls below Earth's radius. This is not an issue for the numbers considered here, but could be for large throw speeds. The eccentricity can be calculated from the angular momentum at launch, $\boldsymbol{L} = R_o\hat{\boldsymbol{x}} \times m\boldsymbol{v}_{SE}$, where $m$ is the mass of the sandwich, which we do not need to specify. Use Equation (7.25) for $\boldsymbol{v}_{SE}$ and Equation (5.10) with $\mu \sim m$ gives the magnitude of $L$ as $L^2 = GMm^2a(1 - e^2)$. You will also need $GM = v_o^2 R_o$; the eccentricity emerges as

$$\varepsilon = \sqrt{1 - \left(\frac{R_o}{a}\right)\left[1 + \left(\frac{v_{SB}}{v_o}\right)\sin\theta_{SB}\right]^2}. \tag{7.28}$$

For the situations considered here, all $\varepsilon$ values are <0.01.

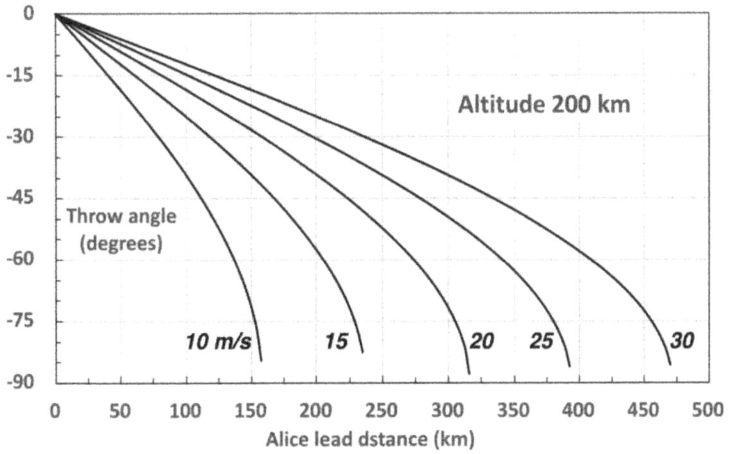


**Figure 7.5.** Bob's throw angle as a function of Alice's lead distance for throwing speeds of 10, 15, 20, 25, and $30\,\mathrm{m\ s^{-1}}$.

If Alice lags Bob, a nuance arises. In this case we will have $\theta_o < 0$, which gives a semimajor axis $a > R_o$ and a period $> T_0$. Here Alice catches the sandwich on her second pass through the launch point, that is, after completing a little more than one orbit. However, the launch speed relative to Earth in this case will be greater than $v_o$, which means that the numerator of Equation (7.26) may be positive or negative, depending on the amount of lag and the magnitudes of $v_o$ and $v_{SB}$. For small lead or lag angles (the only practical cases), we can treat Equation (7.23) with a binomial expansion:

$$a = R_o\,(1 - \theta_o/2\pi)^{2/3} \sim R_o(1 - \theta_o/3\pi), \tag{7.29}$$

where we have used $(1 + x)^n \sim 1 + nx$ for small values of $x$. Using this expression for $a$ in Equation (7.27) gives

$$\begin{aligned} v_{SE}^2 - v_o^2 &= GM\left(\frac{1}{R_o} - \frac{1}{a}\right) \sim GM\left[\frac{1}{R_o} - \frac{1}{R_o(1 - \theta_o/3\pi)}\right] \\ &\sim \frac{GM}{R_o}\left[1 - \frac{1}{(1 - \theta_o/3\pi)}\right] \sim \frac{GM}{R_o}[1 - (1 + \theta_o/3\pi)] \\ &\sim -\frac{GM}{R_o}\left(\frac{\theta_o}{3\pi}\right) \sim -v_o^2\left(\frac{\theta_o}{3\pi}\right), \end{aligned} \tag{7.30}$$

where in the middle step we used $1/(1 - x) \sim 1 + x$ and in the last step we set $GM/R_o = v_o^2$. Remember that in this formulation, $\theta_o > 0$ corresponds to Alice leading Bob.

With this, Equation (7.26) becomes

$$\sin\theta_{SB} \sim \frac{-v_o^2(\theta_o/3\pi) - v_{SB}^2}{2\,v_o\,v_{SB}}. \tag{7.31}$$

If her lag is very slight ($\theta_o < 0$, but small in magnitude compared to $v_{SB}^2$), the numerator will be negative and Bob will still have to make a backwards throw; once the lag exceeds $|\theta_o| > 3\pi v_{SB}^2/v_o^2$, he must throw forward into the first or second quadrant.

A final comment on Figure 7.5. For a given throw speed, it is clear from the behavior of the curves that Alice's lead distance cannot exceed some limiting value, which occurs for a directly backwards throw, that is, one with $\theta_{SB} = -90°$. What is the relationship between $v_{SB}$ and Alice's lead distance for such throws? We can determine this from Equations (7.26) and (7.30); there will be no harm in assuming that the lead angle is small. For $\theta_{SB} = -90°$, $\sin\theta_{SB} = -1$, and Equation (7.26) can be written as

$$v_{SE}^2 - v_o^2 = v_{SB}^2 - 2\, v_o\, v_{SB}. \tag{7.32}$$

Now, from the numbers we have dealt with above, we know $v_{SB} \sim 30$ m s$^{-1}$ and $v_o \sim 7800$ m s$^{-1}$. On the right side of this expression, we can safely assert that $v_{SB}^2 \ll -2\, v_o\, v_{SB}$ and hence drop the $v_{SB}^2$ term. For the left side, invoke Equation (7.30). Then cancel a factor of $v_o$ in what remains to give

$$v_{SB} \sim \frac{\theta_o}{6\pi} v_o \sim \frac{\theta_o}{6\pi}\sqrt{\frac{GM}{R_o}}. \tag{7.33}$$

If Alice's lead is $s$ meters, then $s = R_o\theta_o$. Use this to eliminate $\theta_o$ gives $v_{SB}$ in terms of the lead distance:

$$v_{SB} \sim \frac{s}{6\pi R_o}\sqrt{\frac{GM}{R_o}} \qquad (\theta_{SB} = -90°). \tag{7.34}$$

This expression gives results in excellent accord with Figure 7.5. A 500 km lead in an orbit of radius 6570 km corresponds to a lead angle of only about 0.7 degrees, so the small-angle approximation is well-satisfied. If Alice's lead is such that the right side of (7.34) exceeds Bob's maximum throw speed, they cannot make the transfer.

In the end, it would just be easier to sneak a sandwich aboard before launch. A few years after DuBridge's' talk, Gemini 3 astronaut John Young did exactly this, albeit with a corned beef sandwich. The experiment at in-flight dining was not particularly successful, however, as crumbs of rye bread began to float around the cabin and concern arose that they could interfere with spacecraft operations https://www.space.com/39341-john-young-smuggled-corned-beef-space.html.

## References

https://en.wikipedia.org/wiki/Lambert%27s_problem

DuBridge, L. A. 1960, AmJPh, 28, 719

https://www.youtube.com/watch?v=Ky2XIElijvs

https://www.space.com/39341-john-young-smuggled-corned-beef-space.html

# Chapter 8

## Some Sundry Results

This chapter collects a few miscellaneous results on (mostly) elliptical orbits. Section 8.1 examines the question of how to compute the "average" distance of a planet from the Sun. It turns out that there is more than one way of determining this; we will look at four methods of varying complexity. Section 8.2 considers an analogous question, the average speed of a planet in an elliptical orbit. Section 8.3 addresses how to get a satellite into an orbit of a desired orientation, semimajor axis, and eccentricity when it is "launched" from a given distance from the gravitating body; this is akin to the analysis of the Lambert problem in Section 7.2 except that no time constraint is involved. Section 8.4 examines how the gravitational forces of the Earth and Moon are used in combination to keep the *James Webb Space Telescope* in its distant orbit. Section 8.5 deals with an approximate treatment of a phenomenon known as the perihelion advance of Mercury. This was a significant issue in the history of physics in that it took Einstein's General Theory of Relativity to arrive at a full explanation; our approximate treatment is based on a property of modified Keplerian ellipses that can be contrived to reproduce the relativistic result, admittedly with some "knowing the answer." Section 8.6 considers an example of how many of the expressions that we have encountered that can be transformed into more convenient units. Section 8.7 describes how we can deduce the orientation of Earth's orbit in the sense of determining where the spring equinox occurs relative to perihelion and aphelion. Section 8.8 considers the motion of the Sun about the center of mass of the solar system. Section 8.9 considers the issue of using a planet to create a "gravitational slingshot" to redirect the trajectory of a space probe. Section 8.10 offers a few concluding words.

### 8.1 Average Distance of a Planet from the Sun

Asking what is the average distance of a planet from the Sun seems a simple enough question. You might then be surprised to learn that four such averages can be found in astronomy textbooks and technical literature! One of these is very simple

doi:10.1088/2514-3433/acb430ch8

(and misleading), while the other three are more technically complex. Even more surprisingly, the simplest method gives exactly the same result as one of the more sophisticated approaches.

Why are there different approaches to computing this average? With the exception of the simplest method, which is treated in Section 8.1.1, the ambiguity lies in the fact that we can take the average with respect to different variables. In each case, the distance $r$ is expressed as a function of the desired variable, and an integration is carried out over the range of that variable. In other words, one has to specify "average with respect to *what?*" To make a loose analogy, if you have a spreadsheet of census information, you could compute an "average" person on the basis of height, weight, age, income, number of children, numbers of automobiles owned, number of years of education, and so on. They all have some meaning.

To formalize this, suppose that $r$ can be expressed as a function of some independent variable $\lambda$, $r(\lambda)$. Let $\Lambda$ be the integral of $\lambda$ over one orbit,

$$\Lambda = \int_{\text{orbit}} d\lambda. \tag{8.1}$$

The average of $r(\lambda)$ over an orbit is then

$$\langle r \rangle_\lambda = \frac{1}{\Lambda} \int_{\text{orbit}} r(\lambda) d\lambda. \tag{8.2}$$

We will look at the cases of $\lambda$ being time $t$, the apsidal angle $\phi$, and elements of arc-length $ds$ of the elliptical orbit; this latter method connects to the simple, non-integral method. In practice, since elliptical orbits are symmetric about their major axes, we ultimately need to integrate over only half an orbit. The corresponding values of $\Lambda$ are then one-half of the orbital period ($T/2$), $\pi$ radians, and $S/2$, where $S$ is the perimeter of an ellipse. Which is the "right" average? That is a matter of taste —but be sure to specify your taste.

Before proceeding, there is an operational issue to be discussed here. Ultimately, we know the position vector explicitly only as a function of $\phi$ or $\psi$. If $\lambda$ is, say, the time $t$, we first have to formulate $dt$ in terms of $d\phi$ or $d\psi$. Ultimately, all of our integrals will be over an angular argument, but the integrands are different according as the transformations involved. Think of these averages as weighting $r$ by an element of the averaging variable involved.

### 8.1.1 The Simplest Average

If you want a shortcut way to the average distance, write down the aphelion and perihelion distances $a(1 + \varepsilon)$ and $a(1 - \varepsilon)$, add them together to get $2a$, and divide by 2 to get $\langle r \rangle = a$. Thus, you might read—colloquially—that the semimajor axis *is* the average distance.

A few minutes' reflection on this approach reveals it to be quite unsatisfying, however: there are many possible distances between $a(1 + \varepsilon)$ and $a(1 - \varepsilon)$, and the planet has a different speed at each; the values at perihelion and aphelion are the extremes. This approach takes no account of this variation, and is why other methods have been developed. Historically, the Astronomical Unit was defined in this way, but since 2012 the International Astronomical Union has defined it to be exactly 149,597,870,700 m.

### 8.1.2 The Time-Averaged Distance

This is probably the most intuitively appealing measure of $\langle r \rangle$: imagine dividing the orbit into increments of time, determine $r(t)$ for each, add up all of the $r(t)$ values, and divide by the elapsed time; designate the result as $\langle r \rangle_t$. In the language of Equations (8.1) and (8.2),

$$\langle r \rangle_t = \frac{2}{T} \int_{\text{half-orbit}} r(t) dt. \tag{8.3}$$

Now, the problem with this was alluded to above: we know $r(\phi)$ from Equation (4.3), not $r(t)$:

$$r = \frac{a(1 - \varepsilon^2)}{1 - \varepsilon \cos \phi}. \tag{8.4}$$

However, our trusty friend Equation (3.36) again comes to the rescue by giving us a way to convert $dt$ to $d\phi$:

$$dt = \frac{\mu}{L} r^2 d\phi. \tag{8.5}$$

With this, Equation (8.3) becomes

$$\langle r \rangle_t = \frac{\mu}{LT} \int_0^{2\pi} r^3 d\phi = \frac{2\mu}{LT} a^3 (1 - \varepsilon^2)^3 \int_0^{\pi} \frac{d\phi}{(1 - \varepsilon \cos \phi)^3}. \tag{8.6}$$

The prefactor can be simplified (check it!) with Equations (5.10), (5.42), and the definition of the reduced mass to the more compact form

$$\frac{2\mu}{LT} a^3 (1 - \varepsilon^2)^3 = \frac{a(1 - \varepsilon^2)^{5/2}}{\pi}. \tag{8.7}$$

The integral in Equation (8.6) is exactly that of Equation (1.11), which gives

$$\langle r \rangle_t = \frac{a(1 - \varepsilon^2)^{5/2}}{\pi} \frac{\pi(1 + \varepsilon^2/2)}{(1 - \varepsilon^2)^{5/2}} = a(1 + \varepsilon^2/2). \tag{8.8}$$

A higher-eccentricity orbit has a greater $\langle r \rangle_t$: The planet spends more time out at aphelion than in the vicinity of the Sun at perihelion.

**Exercise:** At what value of $\phi$ is the radial distance $r$ equal to $\langle r \rangle_t$?

Answer: $\cos \phi = 3\varepsilon/(2 + \varepsilon^2)$.

**Exercise:** Use Equation (1.16) to show that $\langle r^n \rangle_t = a^n (1 - \varepsilon^2)^{(n+1)/2} P_{n+1}(x)$, where the argument $x$ is as in (1.17). Similarly, use (1.19) to show that $\langle 1/r^n \rangle_t = P_{n-2}(x)/a^n(1 - \varepsilon^2)^{(n-1)/2}$.

### 8.1.3 The Angle-Averaged Distance

Another way to compute the average distance is by averaging over the apsidal angle $\phi$, that is, to compute $\langle r \rangle_\phi$. This is actually easier than computing $\langle r \rangle_t$ as no change of variable is involved; integral (1.19) is helpful:

$$\begin{aligned}\langle r \rangle_\phi &= \frac{1}{\pi}\int_0^\pi r(\phi)d\phi = \frac{a(1-\varepsilon^2)}{\pi}\int_0^\pi \frac{d\phi}{1-\varepsilon\cos\phi} \\ &= \frac{a(1-\varepsilon^2)}{\pi}\frac{\pi}{(1-\varepsilon^2)^{1/2}} = a\sqrt{1-\varepsilon^2}.\end{aligned} \tag{8.9}$$

In contrast to $\langle r \rangle_t$, $\langle r \rangle_\phi$ *decreases* with increasing eccentricity; as $\varepsilon \to 1$, $\langle r \rangle_\phi \to 0$. To get a qualitative feel for this, sketch two orbits, one of low eccentricity and one of high. Divide them into steps of equal angle, and consider how $r(\phi)$ behaves when you compare the two.

### 8.1.4 The Arc Length-Averaged Distance

The last method of average distance we consider is to imagine chopping up an elliptical trajectory into increments of path-length $ds$ and forming the average distance as

$$\langle r \rangle_S = \frac{\int r\, ds}{\int ds} = \frac{\int r\, ds}{S}, \tag{8.10}$$

where the symbol $S$ is used to designate the length of the periphery of an ellipse: $S = \int ds$.

This may seem like a strange way to define the average distance (and not easy for astronomers to measure!), but we will see that it leads to an interesting result. The concept is sketched in Figure 8.1: the orbit is divided into infinitesimal steps of length $ds$, each with its own distance $r$ from the Sun. Each segment length is "weighted" by its particular $r$ value, and we want to sum up all of the weighted lengths $rds$ around the ellipse.

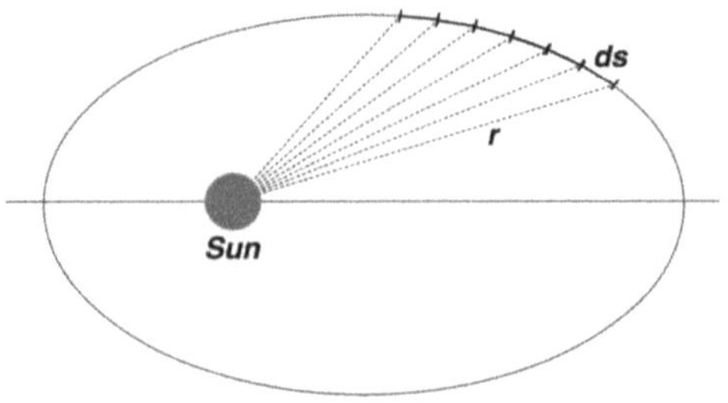


**Figure 8.1.** Concept of arc-length segments. Each segment is oriented to be tangent to the ellipse.

How do we form an element of path length? After all, the orientation of each segment will depend on the value of the apsidal angle of its position; we will have to integrate over a positional angle, getting an expression for *ds* in terms of $\phi$.

We appeal to some basic physics. If an orbiter is moving at speed $v$ and we let time *dt* elapse, then it will move through a distance $ds = v\, dt$. We have an expression for $v$ from Equation (5.15):

$$v = \sqrt{\frac{G(M+m)}{a(1-\varepsilon^2)}(1+\varepsilon^2 - 2\varepsilon\cos\phi)}. \tag{8.11}$$

*dt* can be converted to $d\phi$ through Equation (3.36):

$$dt = \frac{\mu r^2}{L} d\phi. \tag{8.12}$$

This leads to quite a mess:

$$ds = v\, dt = \frac{\mu a^2(1-\varepsilon^2)^2}{L}\sqrt{\frac{G(M+m)}{a(1-\varepsilon^2)}}\frac{\sqrt{1+\varepsilon^2-2\varepsilon\cos\phi}}{(1-\varepsilon\cos\phi)^2}d\phi. \tag{8.13}$$

Attempting to integrate *ds* or *r ds* would be hopeless; they cannot be done in closed form, and even attempting to make some sort of binomial expansion on $\varepsilon$ would be daunting.

To avoid this, it is much cleaner to formulate *ds* in terms of the eccentric anomaly angle $\psi$ of Chapter 6 and invoke the standard calculus-text way of calculating an arc length:

$$ds = \sqrt{dx^2 + dy^2}. \tag{8.14}$$

Look back at Figure 6.1. We approach this by expressing the $(x, y)$ position of the planet in terms of $a$, $\varepsilon$, and $\psi$, and then deriving an expression for *ds* using Equation (8.14).

An expression for $x$ is easy. From Figure 6.1 you should be able to see that

$$x = a\cos\psi. \tag{8.15}$$

$y$ is a little more involved. You should also be able to see that $y = r\sin\phi$. But we can then use Equations (6.4) and (6.7) to get $r$ and $\sin\phi$ in terms of $a$ and $\psi$:

$$y = r\sin\phi = a(1+\varepsilon\cos\psi)\left[\frac{\sqrt{1-\varepsilon^2}\sin\psi}{(1+\varepsilon\cos\psi)}\right] = a\sqrt{1-\varepsilon^2}\sin\psi. \tag{8.16}$$

Taking differentials of these expressions gives:

$$dx = -a\sin\psi\, d\psi \tag{8.17}$$

and

$$dy = a\sqrt{1-\varepsilon^2}\cos\psi\, d\psi. \tag{8.18}$$

Gathering Equations (8.17) and (8.18) into Equation (8.14) gives

$$\begin{aligned} ds &= \sqrt{dx^2 + dy^2} \\ &= \sqrt{(-a \sin\psi \, d\psi)^2 + \left(a\sqrt{1-\varepsilon^2} \cos\psi \, d\psi\right)^2} \\ &= a\sqrt{1-\varepsilon^2 \cos^2\psi} \, d\psi. \end{aligned} \tag{8.19}$$

This is a *much* more compact expression to work with than Equation (8.13). However, despite its innocent appearance, Equation (8.19) cannot be integrated to give a closed-form expression for the perimeter of an ellipse; $\int ds$ is what mathematicians refer to as, for obvious reasons, an "elliptic" integral. But for our purpose of computing $\langle r \rangle_S$, it turns out that we don't in fact have to try to compute $\int ds$ in Equation (8.10). Why then did I go to the trouble of setting up Equation (8.19)? The answer is that it will prove very handy in the following section.

To get to the average $\langle r \rangle_S$, combine Equations (6.4) and (8.10):

$$\begin{aligned} \langle r \rangle_S &= \frac{\int r \, ds}{\int ds} = \frac{\int a(1 + \varepsilon \cos\psi) \, ds}{S} \\ &= \frac{a\int ds + a\int \varepsilon \cos\psi \, ds}{S} = \frac{aS + a\int \varepsilon \cos\psi \, ds}{S} \\ &= a + \frac{a\varepsilon}{S} \int \cos\psi \, ds. \end{aligned} \tag{8.20}$$

Consider the integral appearing here. From the symmetry of an ellipse about its major axis, we can integrate $\psi$ over the range 0 to $\pi$ and multiply the result by two:

$$\frac{2a\varepsilon}{S} \int_0^{\pi} \cos\psi \, ds. \tag{8.21}$$

Here is the key point: $\cos\psi$ is an antisymmetric function about $\psi = \pi/2$ in that it is positive between 0 and $\pi/2$, and negative between $\pi/2$ and $\pi$, that is, for some argument $w$, $\cos(\pi/2 + w) = -\cos(\pi/2 - w)$. *The integral in equation (8.21) will be exactly equal to zero by virtue of this asymmetry.*

What remains of Equation (8.20) is just the first term:

$$\langle r \rangle_S = a. \tag{8.22}$$

Remarkably, this is exactly the same result as given by the shortcut method in Section 8.1.1. Authors who adopt the shortcut method can make a hand-waving appeal to "a more rigorous calculation," although they should specify what quantity the average is being taken with respect to—and be honest that the idea of equal-length path segments is not likely to be the first averaging method that comes to mind.

## 8.2 Time-Average Orbital Speed

A question that may have occurred to you while reading the preceding section is to ask what is the time-average speed $\langle v \rangle_t$ of a planet in an elliptical orbit. This can be computed from the equation for an element of arc length along an ellipse, Equation (8.19). The basic idea is to integrate this expression halfway around the orbit, that is, for $\psi$ running over the range $(0, \pi)$, multiply by 2 to get the total perimeter, and then divide by the orbital period $T$:

$$\langle v \rangle_t = \frac{2a}{T} \int_0^{\pi} \sqrt{1 - \varepsilon^2 \cos^2 \psi}\, d\psi. \tag{8.23}$$

Now, as remarked in the preceding section, this integral cannot be done in closed form. Rather, we will have to settle for a series expansion in terms of increasing powers of the eccentricity. I will carry out the derivation to terms in sixth order, that is, an expansion that includes terms up to $\varepsilon^6$.

The key concept here is the technique of a binomial expansion. We will in particular use the expansion (1.22) from Chapter 1:

$$\sqrt{1 \pm x} = 1 \pm \frac{1}{2}x - \frac{1}{8}x^2 \pm \frac{1}{16}x^3 - \frac{5}{128}x^4 + \cdots (x^2 < 1). \tag{8.24}$$

I won't need the $x^4$ term here, but it will be useful for an exercise at the end of this section. We have $x = \varepsilon^2 \cos^2 \psi$, which gives

$$\begin{aligned} \langle v \rangle_t = \frac{2a}{T}\Bigg[ &\int_0^{\pi} d\psi - \frac{\varepsilon^2}{2}\int_0^{\pi} \cos^2 \psi\, d\psi - \frac{\varepsilon^4}{8}\int_0^{\pi} \cos^4 \psi\, d\psi \\ &- \frac{\varepsilon^6}{16}\int_0^{\pi} \cos^6 \psi\, d\psi + \cdots \Bigg]. \end{aligned} \tag{8.25}$$

Notice that only even powers of $\varepsilon$ are involved.

Now, it turns out that there is an extremely handy integral for even powers of $\cos \psi$ over the range $(0, \pi)$:

$$\int_0^{\pi} \cos^{2m} d\psi = \pi \left[ \frac{(2m)!}{2^{2m}(m!)^2} \right], \tag{8.26}$$

where the exclamation marks designate factorials.

With this identity, the integrals in Equation (8.25) respectively evaluate to $\pi$, $\pi/2$, $3\pi/8$, and $5\pi/16$. Incorporating these results gives

$$\langle v \rangle_t = \frac{2\pi a}{T}\left(1 - \frac{1}{4}\varepsilon^2 - \frac{3}{64}\varepsilon^4 - \frac{5}{256}\varepsilon^6 + \cdots\right). \tag{8.27}$$

If the eccentricity is zero, we have the speed for a circular orbit of radius $a$ and period $T$, namely $2\pi a/T$. For non-zero eccentricity, the average speed is always *less* than that for a circular orbit whose radius is the semimajor axis of the ellipse.

**Exercise:** Show that the next term in the expansion for $\langle v \rangle_t$ is $-(175/16{,}384)\varepsilon^8$.

**Exercise:** In a calculation of average speed, you want the $\varepsilon^6$ term to not exceed 10% of that of the $\varepsilon^4$ term. What is the maximum tolerable $\varepsilon$? Answer: $\varepsilon \sim 0.49$.

To close this section, another exercise along this line is to calculate the time-average of the squared orbital speed, $\langle v^2 \rangle_t$. This is fairly easily to do in terms of the usual apsidal angle $\phi$, using Equation (3.36) to again convert $dt$ to $d\phi$:

$$\langle v^2 \rangle_t = \frac{1}{T}\int_0^T v^2(t)dt = \frac{2\mu}{LT}\int_0^\pi v^2(\phi)r^2 d\phi. \tag{8.28}$$

With Equation (5.15) and the equation for an ellipse, this becomes

$$\langle v^2 \rangle_t = \frac{2\mu}{LT}G(M+m)a(1-\varepsilon^2)\int_0^\pi \frac{(1+\varepsilon^2-2\varepsilon\cos\phi)}{(1-\varepsilon\cos\phi)^2}d\phi. \tag{8.29}$$

This integral can be done with the integrals listed in Section 1.4 and using Equation (5.10) for $L$. You should be able to show that

$$\langle v^2 \rangle_t = \frac{G(M+m)}{a}. \tag{8.30}$$

This is exactly the product of the orbital speeds at aphelion and perihelion. This means that the root-mean-square speed $\sqrt{\langle v^2 \rangle_t}$ is equal to the geometric mean of the speeds at aphelion and perihelion.

**Exercise:** In computing $\langle v^2 \rangle_t$, I switched to integrating over $\phi$, as opposed to integrating over $\psi$ as I did when computing $\langle v \rangle_t$. What would have been the difficulty in writing $v = ds/dt$, using the expression (8.19) for an elliptical arc-length $ds$, setting $v^2 = (ds/dt)^2$, and then computing $\frac{1}{T}\int v^2 dt$? Why didn't I run into a problem when calculating $\langle v \rangle_t$? Sometimes, doing a calculation in one coordinate representation can be much easier than in another.

**Exercise:** Compute the time-averaged value of the potential energy over the course of one orbit: $\langle V \rangle = \frac{1}{T}\int_0^T V(r)dr$. Equation (3.23) will be helpful, and see the integrals in Chapter 1. Then compare your answer to the time-averaged kinetic energy $\langle KE \rangle = \frac{1}{2}\mu\langle v^2 \rangle_t$ using Equation (8.30). You should find $\langle V \rangle = -2\langle KE \rangle$. This can be shown to be a specific case of a more general result known as the virial theorem, which turns up in advanced classical and quantum mechanics analyses of multi-particle systems.

## 8.3 Determining Initial Launch Conditions

Suppose that we have a satellite of mass $m$ which is at some distance $r_{\text{initial}}$ from a gravitating body of mass $M$, with $M \gg m$. We wish to put it into an orbit of semimajor axis $a$ and eccentricity $\varepsilon$ around the body. What initial speed and direction of travel will have to be arranged for? With $M \gg m$, the reduced mass, to a very good approximation, will be $\mu \sim m$. Presumably we can control the value of $r_{\text{initial}}$ by having the satellite carried on a transport rocket to the insertion point, as sketched in Figure 8.2. Given $r_{\text{initial}}$, the required initial speed is dictated by

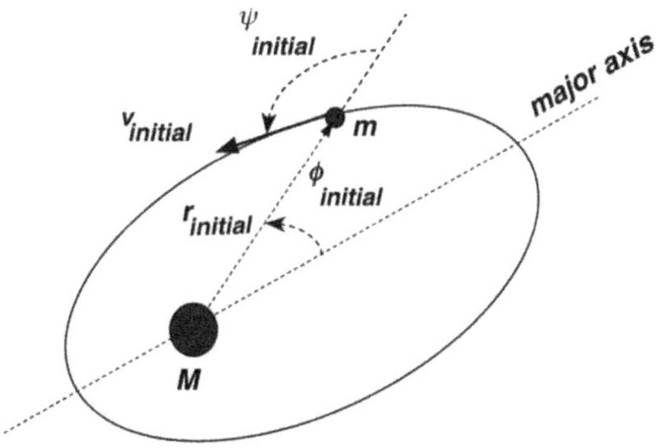


**Figure 8.2.** Initial conditions for satellite orbital insertion.

conservation of energy and the desired semimajor axis by combining Equations (5.19) and (2.9), with the latter in $\mu v^2/2$ form:

$$E = -\frac{GMm}{2a} = \frac{1}{2}\mu v_{\text{initial}}^2 - \frac{GMm}{r_{\text{initial}}}. \tag{8.31}$$

The desired eccentricity fixes the total angular momentum of the system via Equation (5.10) with $\mu \sim m$:

$$L^2 = GM\, m^2 a(1 - \varepsilon^2). \tag{8.32}$$

The direction of launch can be determined from the fact that angular momentum is a vector quantity, $\boldsymbol{L} = \boldsymbol{r} \times \boldsymbol{p} = \boldsymbol{r} \times m\boldsymbol{v}_{\text{initial}}$. The magnitude of $L$ is then $r_{\text{initial}}\, m\, v_{\text{initial}} \sin \psi$, which can be used to compute the direction $\psi_{\text{initial}}$ relative to the launch-position vector. The value of the initial azimuthal angle in the orbit $\phi_{\text{initial}}$ can then be found from the equation for the elliptical orbit. Note that $\psi$ here is *not* the eccentric anomaly, but rather the angle between the direction of the initial position and velocity vectors.

**Exercise:** A satellite of mass $m = 1000$ kg is to be put into an orbit around the Sun with a semimajor axis of one AU and eccentricity 0.7 in order that it can sample the solar wind during close passages to the Sun. It is released from a carrier rocket at an initial distance of 0.8 AUs from the Sun. What must be its initial speed and launch angle $\psi_{\text{initial}}$? What will be its initial azimuthal angle $\phi$? Answer: $v_{\text{initial}} = 36.47$ km s$^{-1}$; $\phi_{\text{initial}} = 58.8°$ or 310°.2, with $\psi_{\text{initial}} = 133°.2$ or 46°.8. This is illustrated in Figure 8.3.

## 8.4 The $L_2$ Lagrange Point and the *James Webb Space Telescope*

The *James Webb Space Telescope* (JWST) was launched on 2021 December 25, and with its primary mirror of diameter 6.5 m is the largest optical/infrared telescope in space. JWST has already begun collecting a trove of data on the early universe.

You may have read that JWST was put into an orbit around the *Sun* about 1.5 million km from Earth, an "altitude" far greater than that of most satellites.

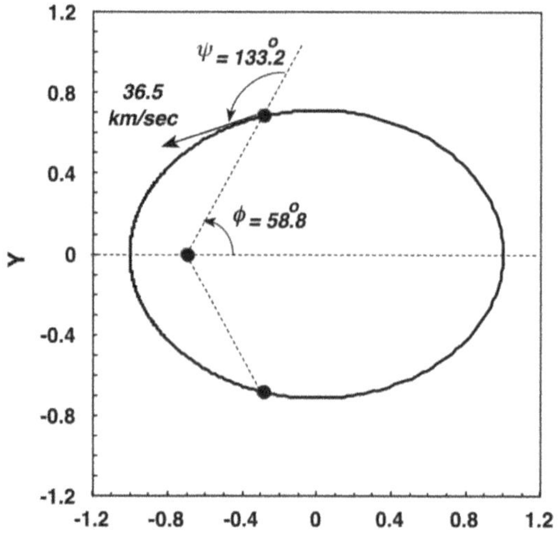


**Figure 8.3.** Plot of an ellipse with semimajor axis $a = 1\,\text{AU}$ and eccentricity $\varepsilon = 0.7$. The dot at $(x, y) = (-0.7, 0)$ is the location of the focus. The two dots on the ellipse are mirror-image initial locations with $r = 0.8\,\text{AU}$.

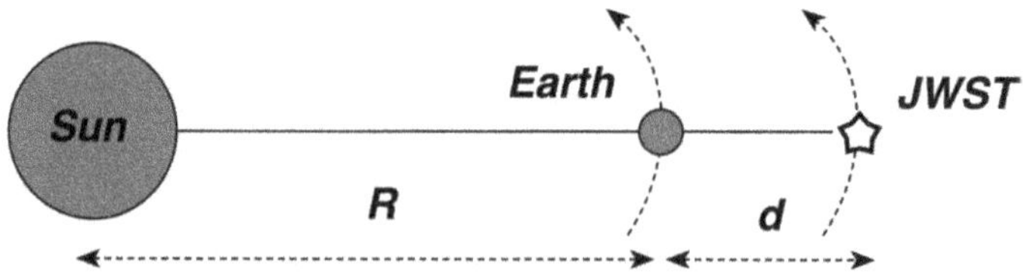


**Figure 8.4.** Sketch of Earth and JWST "co-moving" in their orbits around the Sun. Not to scale.

Specifically, it was put into a location known as the $L_2$ Lagrange point. In general, a Lagrange point is a point of equilibrium of a small mass under the gravitational influence of two much more massive bodies. "Equilibrium" here is not to be interpreted as "not moving," but rather as "stable against outside perturbations" as in Section 5.10. Five Lagrange points, conventionally labeled $L_1$ through $L_5$, are recognized in celestial mechanics; our concern is with the second of these, $L_2$. The Lagrange points are named after the famous Italian/French mathematician Joseph-Louis Lagrange (1736–1813), whose name is attached to many areas of mathematics, astronomy, and physics; see https://en.wikipedia.org/wiki/Lagrangepoint.

A great advantage of the $L_2$ point is that it allows mission planners to use the Earth as a natural Sun-shade. The situation is sketched in Figure 8.4. $R$ is the radius of the Earth's orbit around the Sun, and $R + d$ is the radius of JWST's orbit. It is worth emphasizing that the telescope orbits the Sun, not Earth. The premise is that by arranging the orbital period of JWST to be exactly equal to that of Earth, then Sun–Earth–JWST remain in a line, facilitating the constant shading effect. Think of the $L_2$ point as "co-moving" with Earth.

Now, a planet with a greater orbital radius around the Sun normally has a greater period: Kepler's third law. How then is the co-motion arranged? The trick is that JWST is subjected to the combined gravitational pulls of the Sun and the Earth, which provides for a greater centripetal force than would be present if it were under the influence of the Sun alone. By choosing the Earth–JWST distance $d$ to be just right, this centripetal force provides for just the right period. Our goal is to find an expression for $d$ in terms of $R$ and the masses of the Sun and Earth.

The analysis here is restricted to circular orbits for simplicity. I am grateful to Dr. John Gibson for suggesting this derivation.

For a planet of mass $m$ in a circular orbit of radius $r$ about some force center, we know that it must be subject to a centripetal force

$$F_{\text{cent}} = \frac{mv^2}{r}.$$

It will be helpful to have this in terms of the orbital period $T$, for which we must have $v = 2\pi r/T$:

$$F_{\text{cent}} = \frac{mv^2}{r} = \frac{m}{r}\left(\frac{4\pi^2 r^2}{T^2}\right) = \frac{4\,m\pi^2 r}{T^2}. \tag{8.33}$$

In this scenario, the centripetal force on JWST (mass $M_J$) is provided by its combined gravitational attraction to the Sun (mass $M_S$) and Earth (mass $M_{\text{E}}$):

$$\frac{G\,M_S\,M_J}{(R+d)^2} + \frac{G\,M_{\text{E}}\,M_J}{d^2} = \frac{4\,M_J\,\pi^2(R+d)}{T^2}. \tag{8.34}$$

The factors of $M_J$ can be canceled.

Now, on the right side of Equation (8.34), $T$ is the orbital period of JWST, which we want to be equal to Earth's orbital period. From Kepler's third law applied to Earth, this will be $T^2 = 4\pi^2 R^3/GM_S$, taking the Sun to be much more massive than the Earth. Hence we can write

$$\frac{G\,M_S}{(R+d)^2} + \frac{G\,M_{\text{E}}}{d^2} = 4\,\pi^2(R+d)\frac{G\,M_S}{4\pi^2 R^3}.$$

This can be cleaned up by canceling the factors of $G$ and $4\pi^2$:

$$\frac{M_S}{(R+d)^2} + \frac{M_{\text{E}}}{d^2} = \frac{M_S\,(R+d)}{R^3}. \tag{8.35}$$

We cannot solve for $d$ analytically, but we can invoke a binomial-expansion approach as used in Section 8.2. It turns out (justified later) that $d$ is small compared to $R$, so define

$$x = \frac{d}{R}. \tag{8.36}$$

In Equation (8.35), extract factors of $R$ from within the bracketed terms; when this is done on the right side, you get the combination $R/R^3$, or $1/R^2$. This gives

$$\frac{M_S}{R^2(1+x)^2} + \frac{M_E}{d^2} = \frac{M_S(1+x)}{R^2}. \tag{8.37}$$

Bring the first term on the left side over to the right side:

$$\frac{M_E}{d^2} = \frac{M_S}{R^2}\left[(1+x) - \frac{1}{(1+x)^2}\right]. \tag{8.38}$$

For the second term on the right, invoke the binomial expansion $1/(1+x)^2 \sim 1 - 2x + \cdots$ for small $x$. This gives

$$\frac{M_E}{d^2} \sim \frac{M_S}{R^2}[(1+x) - (1 - 2x + \cdots)] \sim \frac{3M_S}{R^2}x \sim \frac{3M_S}{R^3}d, \tag{8.39}$$

from which we can write

$$\frac{d}{R} \sim \left[\frac{M_E}{3\,M_S}\right]^{1/3}. \tag{8.40}$$

Now, $M_S \sim 333,000\ M_E$, so $d/R \sim 0.010$, that is, the $L_2$ point is about 1% of the Earth–Sun distance from Earth, or about 1.5 million km. For those who prefer more archaic units, this is just under a million miles. The smallness of $d/R$ justifies the binomial expansion.

JWST is far beyond the reach of any crewed repair mission. Should such a mission be necessary, it will have to be robotic.

**Exercise:** What is the angular diameter of Earth as seen from a distance of 1.5 million km? The angular size of the Sun as seen from Earth is about 0.°5 how effective will the shading be? Answer: about 0.°49, practically a perfect eclipse.

## 8.5 An Approximate Treatment of Mercury's Perihelion Advance

In the mid-1800s, Newtonian celestial mechanics experienced a crisis: it was unable to properly predict the orbit of Mercury.

That there was a problem was first pointed out by the French astronomer Urbain Le Verrier (1811–1877), who was closely involved with the discovery of Neptune in 1846; see https://en.wikipedia.org/wiki/UrbainLeVerrier. The problem was an unexplained effect in the "perihelion advance" of Mercury. This is an effect where the return of the planet to perihelion in each orbit does not occur until a slightly later time than would be predicted by the closed orbit of a two-body system. Much of this effect is caused by the gravitational influence of other planets, and so was largely but not completely understood. In the language of astronomical observation, perihelion advances were normally quoted in seconds of arc ("arcseconds," designated with ″) per century. In angular measure, a minute of arc is 1/60 of a degree, and a second of arc 1/60 of that, or 1/3600 of a degree. In the case of Mercury, the predicted Newtonian advance was 531 arcseconds per century. The observed amount, however, was 574 arcseconds per century, a discrepancy of 43 such units. It is powerful testament to the precision of astronomical measurements that such a minute discrepancy could be detected. With its period of 0.2408 years, Mercury completes about 415 orbits per century, so the effect amounts to only about 0.104″ per orbit, or about $5 \times 10^{-7}$ radians per orbit, a number to which we shall return. Nowadays, space-based satellites can measure the positions of stars to accuracies on the order of *micro*arcseconds.

The missing 43″ was not explained until Albert Einstein developed his general theory of relativity in 1915. Einstein's work altered our conceptions of time and space; the missing advance is a manifestation of what is now known as the curvature of spacetime, an effect which has no counterpart in Newtonian physics. We can, however, modify our analysis of Kepler orbits to create an effect which mimics the Einsteinian solution. This is by no means a rigorous treatment; the intent is to give you an idea of what such orbits look like and to further explore some of the orders of magnitude involved.

The modification consists of altering the equation of an ellipse to the form

$$r = \frac{a(1 - \varepsilon^2)}{1 - \varepsilon \cos(k\phi)}. \tag{8.41}$$

The modification is to include the factor of $k$ in the argument of the cosine. Since the argument must still overall be in radians, $k$ must be dimensionless, and, given the minuteness of the discrepant Mercury effect, presumably not much different from unity.

To see how this affects the shape of an orbit, consider the situation at aphelion. (Aphelions advance as well; the historical convention is to refer to this being a perihelion effect.) Suppose we start our planet at $\phi = 0$ at time-zero; at this moment, $r = a(1 + \varepsilon)$, and $\cos(k\phi) = 1$. Now, normally (when $k = 1$), the next aphelion would occur when $\cos(\phi)$ returns to unity, that is, when $\phi = 2\pi$. But if $k$ is put in the picture, the next aphelion will occur when $k\phi = 2\pi$, that is, when $\phi = 2\pi/k$. If $k$ is slightly greater than unity, this will occur before $\phi$ has passed through a full $2\pi$ radians; if $k$ is slightly less than unity, aphelion will occur after $\phi$ has passed through a full $2\pi$ radians. $k$ slightly greater (less) than unity will then correspond to an aphelion or perihelion retardation (advance). Formally, we can write the aphelion alteration as $\Delta\phi = 2\pi/k - 2\pi = 2\pi(1/k - 1)$. If we further write $k = 1 + \delta$ where $\delta$ is likely to be small, then $\Delta\phi = 2\pi[1/(1 + \delta) - 1] \sim 2\pi[(1 - \delta) - 1] \sim -2\pi\delta$, where $\delta < 0$ for aphelion advance and $\delta > 0$ for a retardation. In the case of Mercury, the advance of $5 \times 10^{-7}$ radians per orbit corresponds to $\delta \sim -(5 \times 10^{-7})/2\pi \sim -7.99 \times 10^{-8}$.

Figure 8.5 shows an orbit of the form of Equation (8.41) for $a = 1$, $\varepsilon = 0.25$ and $\delta = 0.05$, that is, $k = 1.05$. This is a large value of $\delta$, but the point is to show the effect. The focus is located at $(x, y) = (0, 0)$. The motion starts at $(x, y) = (1.25, 0)$ and proceeds counterclockwise for five full "orbits" in the sense that $\phi$ runs over $0° \leqslant \phi \leqslant 1800°$. The effect of a non-unity value of $k$ is to cause the orbit to no longer be closed. This value of $\delta$ will cause an aphelion (and perihelion) retardation. This can be seen in Figure 8.6, which shows the value of $r$ as a function of $\phi$; notice how $r$ returns to its extreme values with an angular period of less than 360°.

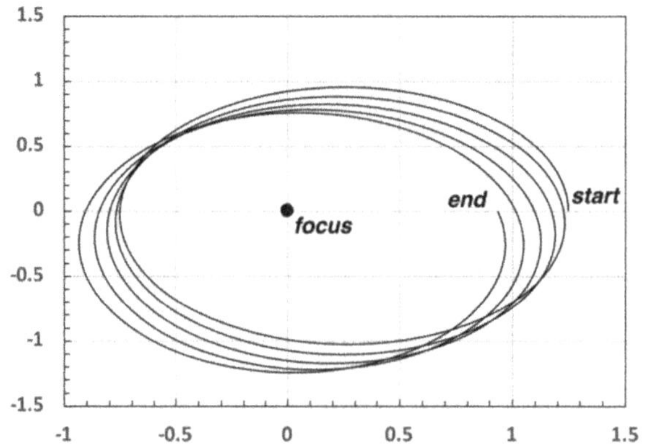


**Figure 8.5.** Modified elliptical orbit of the form of Equation (8.41) with $a = 1$, $\varepsilon = 0.25$, and $k = 1.05$. The focus is at (0,0), and five full "orbits" are shown: $0° \leqslant \phi \leqslant 1800°$.

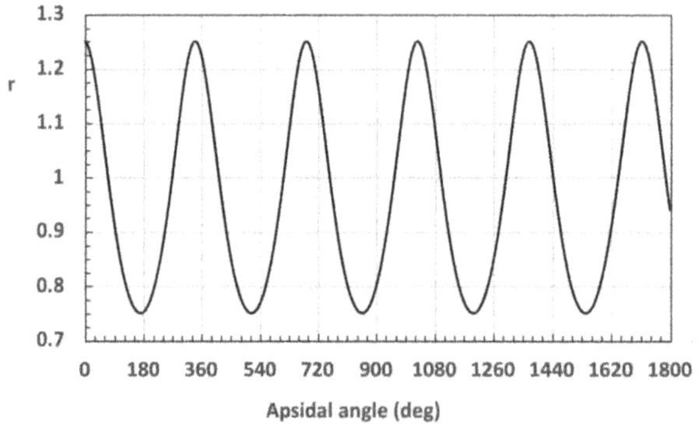


**Figure 8.6.** Behavior of the radial distance $r$ of the orbiter from the focus for the orbit of Figure 8.5. Notice how the return of $r$ to the aphelion (maximum $r$) position recurs after the passage of less than 360° for each orbit; similarly for perihelion.

The effect of $k$ is to introduce an inverse-cube term in the force. From Equation (3.53), the magnitude of the force is found to be (verify!)

$$F(r) = \left[\frac{L^2k^2}{\mu a(1 - \varepsilon^2)}\right]\frac{1}{r^2} + \left[\frac{L^2(1 - k^2)}{\mu}\right]\frac{1}{r^3}. \tag{8.42}$$

If $k = 1$, we recover purely the conventional Newtonian result. If $k$ is slightly different than unity, the Newtonian term will be only slightly altered but the additional inverse-cube term will dominate at very small distances.

If the advance effect can be accommodated by introducing an inverse-cube term, why was the development of relativity necessary? The answer is that other planets also undergo such effects, although to lesser degrees than is the case with Mercury. This requires introducing a different value of $\delta$ for each planet, hardly a "universal" description. In a proper relativistic analysis, Newton's law of gravitation gets modified by the addition of a term which depends on the ratio of a planet's speed to the speed of light, $c$:

$$F(r) = \frac{GMm}{r^2}\left(1 + 6\frac{v^2}{c^2}\right). \tag{8.43}$$

Relativistic effects are usually dependent, to first order, upon the ratio $(v/c)^2$.

## 8.6 A Brief Lesson in Unit Conversion

There is somewhat of a disconnect in celestial mechanics problems in that we usually express physical constants such as $G$ in SI units (meters, kilograms, seconds), which means that our formulae assume the same units. But such units are hardly convenient when dealing with orbital dimensions of possibly thousands or millions of kilometers, orbital periods of anything from hours to centuries, and masses on the order of those of planets or stars. This section illustrates a convenient method for turning a formula into an expression which incorporates units convenient to the of problem at hand.

As an example, if one is using Kepler's third law and considering Earth satellites, it will probably be convenient to express semimajor axes as multiples of Earth's radius, and to express orbital periods in hours.

When the masses involved are such that $M \gg m$, Kepler's third law is

$$T^2 = \left(\frac{4\pi^2}{GM}\right)a^3. \tag{8.44}$$

Now, $G = 6.674 \times 10^{-11}\,\mathrm{m^3\,kg^{-1}\,s^{-2}}$, the radius of the Earth is $R_E = 6.371 \times 10^6$ m, and the mass of the Earth is $M_E = 5.972 \times 10^{24}$ kg. If the orbital period is expressed in hours, ($T_{hr}$) and the semimajor axis in Earth radii as $a_E$, we can write Kepler's third law as (be careful here: "m" means meters, not mass, which being a variable is normally italicized: $m$)

$$(3600\ \mathrm{s}\ T_{hr})^2 = \left[\frac{4\pi^2}{(6.674 \times 10^{-11}\,\mathrm{m^3\,kg^{-1}\,s^{-2}})(5.972 \times 10^{24}\,\mathrm{kg})}\right] (6.371 \times 10^6\ \mathrm{m}\ a_E)^3. \tag{8.45}$$

Note that $T_{hr}$ and $a_E$ will be dimensionless. Rearrange this expression to gather together all the numerical factors:

$$T_{hr}^2 = \left[\frac{4\pi^2(6.371 \times 10^6\,\mathrm{m})^3}{(3600\,\mathrm{s})^2(6.674 \times 10^{-11}\,\mathrm{m^3\,kg^{-1}\,s^{-2}})(5.972 \times 10^{24}\,\mathrm{kg})}\right]a_E^3. \tag{8.46}$$

The term in square brackets evaluates to a dimensionless number:

$$T_{hr}^2 = 1.976\,a_E^3. \tag{8.47}$$

For the purpose of a quick approximation, 1.976 could be rounded to 2 without much harm.

This result is valid for any Earth-orbiting satellite. For instance, if $a_E = 5$ Earth radii, then $T_{hr} = \sqrt{1.976(5)^3} \sim 15.7$ hours. As a rule, you should find that if masses, distances, and times are scaled to units appropriate for the problem at hand, numerical factors such as the 1.976 appearing here will generally be on the order of unity. You might wish to check that for planets orbiting the solar system, the numerical factor is exactly unity if periods are expressed in years and semimajor axes in Astronomical Units—precisely as Kepler himself deduced.

**Exercise:** A "geosynchronous" satellite has a period of 24 hours. What is the semimajor axis of the orbit in Earth radii? Answer: 6.63. Such a satellite would appear to be stationary as seen from a given point on Earth; this is why your satellite dish can maintain a fixed orientation. Such satellites orbit the equator.

## 8.7 Orientation of Earth's Orbit

A common topic in astronomy classes is how the seasons that we experience on Earth are due not to the eccentricity of the orbit causing a constantly changing distance from the Sun, but rather to the fact that the plane of Earth's equator is inclined at about 23.5 degrees to the plane of its orbit; with $\varepsilon = 0.0167$, the distance-change effect is small. The plane of the orbit is known as the ecliptic, which means the path in the sky followed by the Sun over the course of a year. Many texts include an illustration along the lines of Figure 8.7 to show how this inclination causes the northern and southern hemispheres to experience more or less direct sunlight over the course of the year. At two times during the year, the northern-hemisphere fall and spring equinoxes (usually around September 21 and March 21), the Sun is directly overhead at the equator. Around June 21 (northern-hemisphere summer solstice; southern-hemisphere winter solstice) it is overhead at latitude 23°.5 north, and around December 21 (northern-hemisphere winter solstice; southern-hemisphere summer solstice), it is overhead at latitude 23°.5 south. The March 21 equinox is known to astronomers as the vernal equinox (VE), and the location of the Sun against the background sky at that moment serves as a fiducial point for the right-ascension system of stellar coordinates. Northern-hemisphere students are often surprised to learn that the June 21 solstice in fact occurs near the point of orbital aphelion when the Earth is furthest from the Sun,

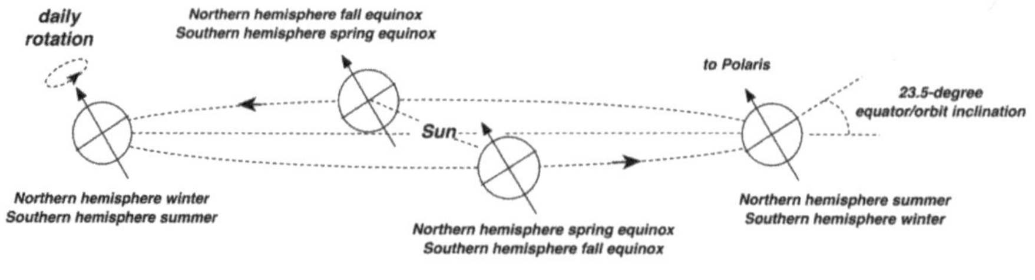


**Figure 8.7.** Sketch of Earth's annual orbit around the Sun.

and conversely that the Christmas time winter solstice occurs near the perihelion point.

Now, the time required to travel from autumnal equinox (AE) through winter solstice (WS) to the vernal equinox (VE), about 179 days, is less than that required to go from vernal equinox through summer solstice (SS) and back to the autumnal equinox, about 186 days. The difference corresponds to only about 2% of the orbital period, so the eccentricity must be slight; also, we can infer that perihelion must occur during the northern winter months, as remarked above.

Often left unaddressed in elementary-level treatments is exactly how we can determine the eccentricity of Earth's orbit and where along the orbit some calibrating point such as a solstice or equinox occurs. In particular, is there any reason from fundamental physics that the northern-hemisphere summer solstice occurs near aphelion, or is this just a coincidence? These issues are examined in this section with an approximate but very accurate method. This material is adapted from a publication by this author Reed (2023).

It is worth remarking that this sort of data can be looked up in any online almanac, where timings will be quoted to an accuracy of a minute. But to adopt such data would be disingenuous: such almanacs are generated by software already programmed with the eccentricity. To keep to the spirit of imagining a Copernicus or Kepler who determines the relevant dates by noting equinoxes as when the Sun rises due east and solstices as those of longest and shortest noontime shadows, I use dates rounded to one day, adopted from my desk calendar. It turns out that even with this one can do reasonably well, determining the eccentricity to an accuracy of about 10%. I do give precise data for comparative purposes, but the intent is to simulate an "eyeball" approach.

Another view of the situation is sketched (not to scale) in Figure 8.8. As usual, the Sun is at the left focus of the orbit and the apsidal angle $\phi$ is measured counter-clockwise from the major axis, with aphelion corresponding to $\phi = 0°$. In this sketch, I have put the summer solstice just somewhat past aphelion. In reality this occurs about two weeks before aphelion, but the diagram is drawn for convenience and this will not affect the following argument.

A key point in this analysis is that solstices and equinoxes are separated by 90° in apsidal angle. This can be argued as follows. Part (b) of Figure 8.8 shows a view in the plane of the ecliptic, perpendicular to the WS–Sun–SS line (*not* perpendicular to the major axis of the ellipse). At summer solstice, the Sun will appear overhead as far north as it gets. To have the same extreme southern latitude at winter solstice to the left side of the sketch, the summer and winter solstice positions must be 180° apart. An identical argument can be made for the positions of vernal equinox and autumnal equinox; they too must be 180° from each other.

Now imagine looking toward the Sun from this external vantage point. The Sun can be over the equator only at the moment when Earth is directly in front of or behind it as viewed from this direction. Thus, the SS–WS and VE–AE lines must be perpendicular, hence the 90° claim. This is a quite general result, independent of any eccentricity of the orbit.

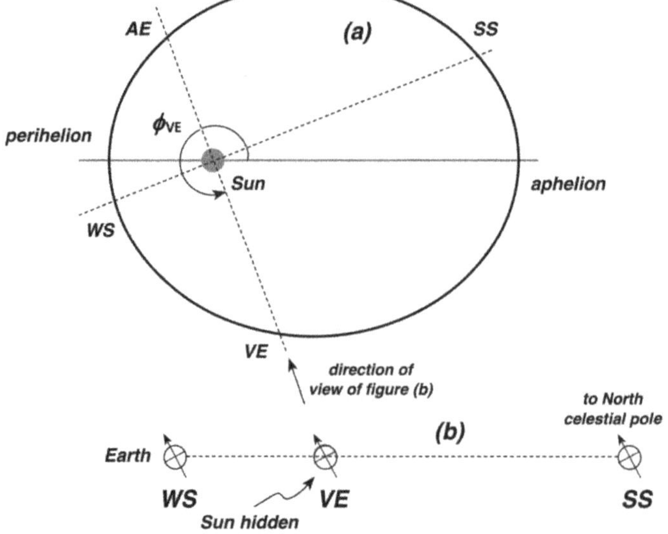


**Figure 8.8.** (a) Sketch of Earth's orbit as seen from above the North pole of the Sun. In this sketch, the eccentricity is $\varepsilon = 0.5$, much greater than the true value of ~0.0167. VE, SS, AE, and WS, respectively, designate vernal equinox, summer solstice, autumnal equinox, and winter solstice. $\phi_{\rm VE}$ is the apsidal angular position of the vernal equinox measured from aphelion. Solstices and equinoxes are 90° apart; their locations here are schematic. (b) View from outside Earth's orbit in the ecliptic plane.

From Equation (5.46), the time for a planet to travel from apsidal angle $\phi_1$ to apsidal angle $\phi_2$ for an orbit of eccentricity $\varepsilon$ is

$$t_{\phi_1 \rightarrow \phi_2} = A \int_{\phi_1}^{\phi_2} \frac{d\phi}{(1 - \varepsilon \cos \phi)^2}, \tag{8.48}$$

where

$$A = \frac{(1 - \varepsilon^2)^{3/2} T}{2\pi}, \tag{8.49}$$

where $T$ is the period of the orbit.

In what follows, ratios of times will be used; it is not necessary to compute or specify $A$.

Now, if $\varepsilon$ is small, the denominator of Equation (8.48) can be treated with a binomial expansion to give

$$t_{\phi_1 \rightarrow \phi_2} \sim A \int_{\phi_1}^{\phi_2} (1 + 2\varepsilon \cos \phi) d\phi,$$

which integrates to

$$t_{\phi_1 \rightarrow \phi_2} \sim A\left[(\phi_2 - \phi_1) + 2\varepsilon(\sin \phi_2 - \sin \phi_1)\right]. \tag{8.50}$$

Now consider the time to travel from VE ($\phi_1$) to SS ($\phi_2$). Since $\phi_{SS} = \phi_{VE} + 90°$, then $\sin\phi_{SS} = \cos\phi_{VE}$, hence $(\phi_2 - \phi_1) = \pi/2$ radians and we have

$$t_{VE\to SS} \sim A\left[\pi/2 + 2\varepsilon(\cos\phi_{VE} - \sin\phi_{VE})\right]. \tag{8.51}$$

Analogous expressions for other 90° spans can be formulated:

$$t_{SS\to AE} \sim A\left[\pi/2 - 2\varepsilon(\sin\phi_{VE} + \cos\phi_{VE})\right]. \tag{8.52}$$

$$t_{AE\to WS} \sim A\left[\pi/2 - 2\varepsilon(\cos\phi_{VE} - \sin\phi_{VE})\right]. \tag{8.53}$$

$$t_{WS\to VE} \sim A\left[\pi/2 + 2\varepsilon(\sin\phi_{VE} + \cos\phi_{VE})\right]. \tag{8.54}$$

There are two unknowns, $\phi_{VE}$ and $\varepsilon$. Taking differences of the times $(t_{VE\to SS} - t_{AE\to WS})$ and dividing by the difference $(t_{WS\to VE} - t_{SS\to AE})$ gives, after some algebra (check it!)

$$\tan\phi_{VE} = \frac{(1 - \tau_1)}{(1 + \tau_1)}, \tag{8.55}$$

where

$$\tau_1 = \left(\frac{t_{VE\to SS} - t_{AE\to WS}}{t_{WS\to VE} - t_{SS\to AE}}\right). \tag{8.56}$$

With $\phi_{VE}$ in hand, $\varepsilon$ can be determined. To avoid dealing with powers and roots, it is again helpful to use a ratio of times to eliminate $A$. Any pair of Equations (8.51)–(8.54) can be used; I use (8.51) and (8.53):

$$\varepsilon \sim \frac{\pi(\tau_2 - 1)}{4(1 + \tau_2)(\cos\phi_{VE} - \sin\phi_{VE})}, \tag{8.57}$$

where

$$\tau_2 = \frac{t_{VE\to SS}}{t_{AE\to WS}}. \tag{8.58}$$

Table 8.1 lists timing data for 2022/23 phenomena from the United States Naval Observatory; these are values predicted by programs pre-loaded with "known" orbital elements. UT designates "universal Time," equivalent to Greenwich Time. The third column of the table lists, to four decimal places, the number of days which have elapsed between successive events; e.g., 92.7368 days pass between vernal equinox and summer solstice. If the orbit were a perfect circle, the same number of days would elapse between each event. The last column lists the number of elapsed *calendar* days, not including the day of the "prior event." These sum to 365, so we can naturally expect some inaccuracy in the results. If you replicate such calculations for other years, beware of subtleties. For example, as measured in universal Time, the 2022 autumnal equinox occurs in the early hours of September 23, but in the Eastern time zone which my calendar keeps, this occurs late in the evening of September 22.

**Table 8.1.** Solstice and Equinox Dates and Times for 2022–2023. From United States Naval Observatory, https://aa.usno.navy.mil/data/EarthSeasons.

| Event | Date/Time (UT) | Days Since Prior Event | Elapsed Calendar Days |
|---|---|---|---|
| Vernal equinox | 2022 March 20 15:33 | — | — |
| Summer solstice | 2022 June 21 09:14 | 92.7368 | 93 |
| Autumnal equinox | 2022 September 23 01:04 | 93.6597 | 93 |
| Winter solstice | 2022 December 21 21:48 | 89.8639 | 90 |
| Vernal equinox | 2023 March 20 21:24 | 88.9833 | 89 |

Using the number of elapsed calendar days, Equation (8.56) gives $\tau_1 = -3/4$, from which (8.55) gives $\phi_{VE} \sim 262°$; Equation (8.57) then gives $\varepsilon \sim 0.0152$. (There will be two solutions for $\tan\phi_{VE}$ separated by 180°, but one of them will give an unphysical negative eccentricity.) The precise USNO numbers give $\phi_{VE} \sim 256°.6$ and $\varepsilon \sim 0.0167$, so the calendar-days estimate of $\varepsilon$ is good to ~9%. Perihelion and aphelion occur about two weeks after winter solstice and summer solstice, around January 4 and July 4, respectively. The sensitivity of results to changes in the intervals can be judged by adding a quarter-day in the SS-to-AE calendar interval (when Earth is moving slowly) to account for the true orbital period; this gives $\phi_{VE} \sim 260°$ and $\varepsilon \sim 0.0158$, which halves the error in the latter.

The precise USNO values place the position of the summer solstice at an orbital position of 13° prior to aphelion, and the corresponding position of the winter solstice at 13° prior to perihelion. The near coincidence of the solstices and extreme distances is just that. Earth's orbit and axial rotation are never the beautifully closed, endlessly-repeating cycles of textbook drawings; we are constantly subject to the gravitational tugs of every object in the solar system. The differential gravitational force of the Sun on the deformable Earth, for example, leads to a precessional motion of the vernal equinox of about 50 seconds of arc per year. This causes the date of perihelion to advance by about one calendar month every 2000 years, although the effect is not uniform. In Newton's time, perihelia occurred in late December as opposed to the present approximately January 4. We have to take the model presented here within the context of these limitations, but it is instructive to see how simple observations served as the foundations of celestial mechanics. Dig out an old calendar and repeat the analysis as a consistency check.

What took Copernicus and Kepler years of manual computational labor can now be accomplished in a class period with a straightforward integral, some timing data, and a hand calculator. Johannes Kepler spent much time defending his mother against accusations of witchcraft. Might he think our computers to be some form of devilry?

## 8.8 Motion of the Sun

In the preceding section we looked at the annual motion of the Earth in its orbit. In this section we take up a similar issue. If we consider the entire solar system to be essentially isolated from the rest of the Galaxy, then its center of mass (CM) must move with constant velocity as seen by an outside observer. For an observer co-moving with the CM, that location will appear stationary even as the Sun and every planet, asteroid, and comet is in motion. (Incidentally, the CM of the solar system is formally known as the "barycenter.") But how dramatic is the Sun's motion? This section investigates this issue with a simple model. I am grateful to Professor Ian Gatland of the Georgia Institute of Technology for suggesting this approach. Interactive simulations along these lines are available online at https://astro.unl.edu/classaction/animations/extrasolarplanets/ca_extrasolarplanets_starwobble.html, https://en.wikipedia.org/wiki/Barycenter.

From Table P.1, it can be seen that well over 99% of the planetary mass of the solar system is wrapped up in the four gas giants, Jupiter, Saturn, Uranus, and Neptune. For the purpose intended here, we need work only with the Sun and these four planets.

To simplify the analysis, I assume that these planets execute circular orbits of radii equal to their semimajor axes about the CM of the solar system, with corresponding periods; the intent is an approximate treatment. The CM is assumed to be at the origin $(x, y) = (0, 0)$ of an "external observer."

Now, the CM of the solar system will be given by (see Equation (3.1))

$$\boldsymbol{r}_{\mathrm{CM}} = \frac{M_{\odot}\boldsymbol{r}_{\odot} + \sum_{\mathrm{planets}} m_i \boldsymbol{r}_i}{M_{\odot} + \sum_{\mathrm{planets}} m_i}, \tag{8.59}$$

where the $\boldsymbol{r}$'s are vector positions relative to the CM, $M_{\odot}$ is the mass of the Sun, and the $m_i$ are the masses of the planets considered. If the CM remains at the origin, then we can set $\boldsymbol{r}_{\mathrm{CM}} = 0$ and solve for the position of the Sun:

$$\boldsymbol{r}_{\odot} = -\left(\frac{\sum_{\mathrm{planets}} m_i \boldsymbol{r}_i}{M_{\odot}}\right). \tag{8.60}$$

Each planet is in motion. If at some fiducial time $t = 0$ a planet is at apsidal angle $\phi_p$, has orbital radius $a_p$, and has angular speed $\omega_p = 2\pi / T_p$, then its position at some later time will be

$$\boldsymbol{r}_p = a_p\left[\cos(\phi_p + \omega_p t)\hat{\boldsymbol{x}} + \sin(\phi_p + \omega_p t)\hat{\boldsymbol{y}}\right]. \tag{8.61}$$

From this, we can compute $\boldsymbol{r}_{\odot}$ as a function of time.

In astronomical almanacs, planetary positions are given in terms of *longitudes* measured from the direction of the vernal equinox. Longitudes for the four gas giants as of 2020 February 1 are listed in Table 8.2. By adopting these as values of $\phi_p$, we get the planets in the correct relative positions as of this date and effectively make the $x$-axis in this analysis to be the direction of the vernal equinox. Values of the semimajor axis and sidereal periods used are those in Table P.1.

Figure 8.9 shows the computed position of the Sun in 0.1-year increments from 2020 to 2080; the scale is in AUs. The greatest extent of the Sun's displacement from

**Table 8.2.** Planetary Longitudes as of 2020 February 1. From Edgar (2020).

| Planet | Longitude (deg) |
|---|---|
| Jupiter | 283.90 |
| Saturn | 295.42 |
| Uranus | 39.08 |
| Neptune | 348.50 |

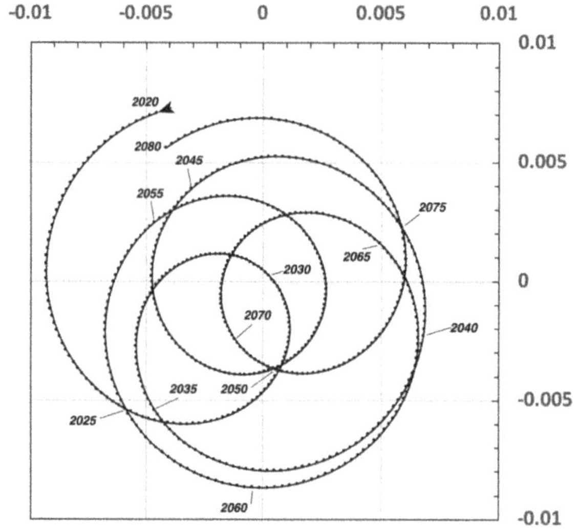


**Figure 8.9.** Motion of Sun, 2020–2080. The scale in both directions is in Astronomical Units. The vernal equinox is in the direction of the $x$-axis.

the CM (to the level of approximation adopted here!) is about 0.01 AU. The diameter of the Sun is ~0.009 AU, so, most of the time, the CM is actually *within* the Sun. This emphasizes that, for all practical purposes, we can consider the Sun to be stationary. If you are reading this around the year 2030, the Sun is now very close to the barycenter of the solar system.

**Exercise:** The full width of Figure 8.9 is 0.02 AU. The star Proxima Centauri is about 268,000 AUs from the solar system. For an observer on Proxima, what would be the angular size, in seconds of arc, of this 0.02 AU width? Ans: About 0.015. This motion would be detectable with current technology—if you are patient.

## 8.9 Gravitational Scattering

This section considers an issue of great relevance to modern solar-system exploration: how to use the gravitational attraction of a planet to redirect the trajectory of a passing space probe. By deftly combining multiple "gravitational scatterings" or "gravitational slingshots," mission planners can arrange for a single probe to pass by two or more target planets.

The situation described here involves an unbound trajectory because the total energy of the probe is positive, so we will not be dealing with a closed elliptical orbit but rather a "fly-by" maneuver.

The situation is shown in Figure 8.10. A probe of mass $m$ is initially moving in the positive $x$-direction at speed $v_o$ toward a planet of mass $M$, which is located at the origin. At the initial location, the two are so far apart that the potential energy $-GMm/r$ is negligible compared to the kinetic energy $mv_o^2/2$, which is consequently the entire energy of the system; this is why the trajectory is unbound. In this simplified treatment, the planet is considered stationary; a proper mission plan would have to account for the motion of the planet.

If the planet's gravitation did not affect the probe, it would continue in a straight line and pass by the planet, missing it by distance $b$ as shown. This distance, not to be confused with the semiminor axis of an elliptical orbit, is known in scattering theory as the "impact parameter." However, the planet attracts the probe, which consequently moves in a hyperbolic arc. As the probe passes the planet, it will speed up because the potential energy $-GMm/r$ becomes appreciable. After the passage, the probe will begin to slow down and will eventually return to its initial speed $v_o$ when it is far from the planet; energy must be conserved.

The point of the analysis is to determine the final direction of travel $\phi_{\text{final}}$ in terms of $M$, $v_o$, $b$, and $G$, that is, the direction of travel when $r \to \infty$. We will also derive expressions for the angular position and distance of the probe at closest approach, its speed at closest approach, and the time it spends in the close vicinity of the planet.

Our analysis will involve both the linear and angular momentum of the probe. Initially, the linear momentum is given by $\boldsymbol{p}_{\text{initial}} = mv_o\hat{\boldsymbol{x}}$. If the probe is initially at $(x, y) = (x_{\text{initial}}, b)$, then its angular momentum about the origin is

$$\boldsymbol{L} = \boldsymbol{r} \times \boldsymbol{p} = (x_{\text{initial}}\hat{\boldsymbol{x}} + b\hat{\boldsymbol{y}}) \times (mv_o\hat{\boldsymbol{x}}) = -(bmv_o)\hat{\boldsymbol{z}}. \tag{8.62}$$

Notice that $\boldsymbol{L}$ does not depend on $x_{\text{initial}}$ because $\hat{\boldsymbol{x}} \times \hat{\boldsymbol{x}} = 0$. The angular momentum is negative because the probe is essentially executing a counterclockwise

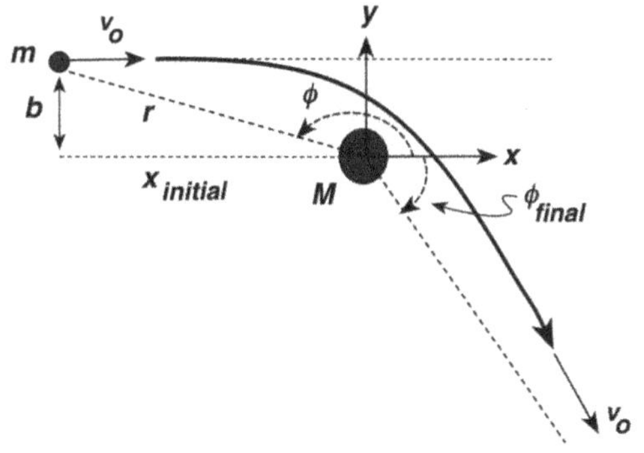


**Figure 8.10.** Geometry of gravitational scattering. The large mass $M$ is assumed to be stationary.

motion around the planet. Because the planet exerts a central force on the probe, $\boldsymbol{L}$ must remain constant in magnitude and direction.

Now, a rigorous analysis of the trajectory would involve integrating Equation (3.42) to determine $\phi(r)$ and then examining what happens as $r$ becomes large. Details on such an approach this can be found in a separate publication by this author and references therein (Reed 2014). However, it turns out that in this situation a simpler approach utilizing a direct application of Newton's second law can also be used, and this is what is done here. As a challenge, you might want to try reproducing the results of this section by integrating Equation (3.42).

In momentum form, Newton's second law reads

$$\frac{d\boldsymbol{p}}{dt} = \boldsymbol{F}. \tag{8.63}$$

Separate variables to write this as $d\boldsymbol{p} = \boldsymbol{F}dt$. Invoke the gravitational force $\boldsymbol{F} = -(GMm/r^2)\hat{\boldsymbol{r}}$, and use Equation (3.36) to replace $dt$ with $(mr^2/L)d\phi$. The factors of $r^2$ cancel; what remains is

$$d\boldsymbol{p} = -\frac{GMm^2}{L}\,\hat{\boldsymbol{r}}\,d\phi. \tag{8.64}$$

Now, as the probe swings around the planet, $\hat{\boldsymbol{r}}$ will be a function of $\phi$. We can deal with this by putting $\hat{\boldsymbol{r}}$ in terms of the constant Cartesian $\hat{\boldsymbol{x}}$ and $\hat{\boldsymbol{y}}$ unit vectors via Equation (1.3):

$$d\boldsymbol{p} = -\frac{GMm^2}{L}\,[(\cos\phi)\hat{\boldsymbol{x}} + (\sin\phi)\hat{\boldsymbol{y}}]\,d\phi. \tag{8.65}$$

Now integrate from the initial circumstance to a later general value of $\phi$. Initially, if the probe is far to the left of the planet, $\phi \sim \pi$, hence

$$\int_{\boldsymbol{p}_{\text{initial}}}^{\boldsymbol{p}(\phi)} d\boldsymbol{p} = -\frac{GMm^2}{L}\int_{\pi}^{\phi}[(\cos\phi)\hat{\boldsymbol{x}} + (\sin\phi)\hat{\boldsymbol{y}}]\,d\phi, \tag{8.66}$$

which evaluates to

$$\boldsymbol{p}(\phi) - \boldsymbol{p}_{\text{initial}} = -\frac{GMm^2}{L}[(\sin\phi)\hat{\boldsymbol{x}} - (\cos\phi + 1)\hat{\boldsymbol{y}}]. \tag{8.67}$$

Now put $L = -(bmv_{\circ})$ from Equation (8.62) and invoke $\boldsymbol{p}_{\text{initial}} = mv_{\circ}\hat{\boldsymbol{x}}$. After some algebra, the later momentum emerges as

$$\boldsymbol{p}(\phi) = mv_{\circ}[(1 + \gamma\sin\phi)\hat{\boldsymbol{x}} + \gamma\,(1 + \cos\phi)\hat{\boldsymbol{y}}], \tag{8.68}$$

where $\gamma$ is a dimensionless (verify this!) combination of the parameters of the problem:

$$\gamma = \frac{GM}{bv_{\circ}^2}. \tag{8.69}$$

Equation (8.68) gives us the momentum of the probe at some general angle $\phi$. We can turn this into an expression for the trajectory $r(\phi)$ by invoking an energy argument. The square of the magnitude of the momentum at any general angle $\phi$ must be the sum of the squares of its components, $p_x^2 + p_y^2$; dividing by $m^2$ then gives the square of the speed as

$$v^2 = v_\circ^2[1 + 2\gamma^2 + 2\gamma(\sin\phi + \gamma\cos\phi)]. \tag{8.70}$$

Now, energy must be conserved:

$$\frac{m}{2}v_\circ^2 = \frac{m}{2}v^2 - \frac{GMm}{r}. \tag{8.71}$$

Cancel the $m$'s in this expression, and use it to eliminate $v^2$ in (8.70). What remains can be reduced to

$$r = \frac{b}{(\gamma + \sin\phi + \gamma\cos\phi)}. \tag{8.72}$$

This is the trajectory of the probe as a function of $\phi$. Notice that for $\phi \to \pi$, we recover the initial condition $r \to \infty$.

To investigate what happens for $r \to \infty$, we demand that the denominator of (8.72) go to zero:

$$\gamma + \sin\phi + \gamma\cos\phi = 0. \tag{8.73}$$

This solution of this expression for $\phi$ in terms of $\gamma$ must give $\phi_{\text{final}}$. This expression can be simplified with the help of two trigonometric identities. These are

$$\sin(2\alpha) = \frac{2\tan\alpha}{1 + \tan^2\alpha} \tag{8.74}$$

and

$$\cos(2\alpha) = \frac{1 - \tan^2\alpha}{1 + \tan^2\alpha}. \tag{8.75}$$

Define $\alpha = \phi/2$ and invoke these in (8.73). Multiply through by $1 + \tan^2\alpha$ to clear out the denominators, and you will find that what remains is $\tan\alpha = -\gamma$, or

$$\phi_{\text{final}} = 2\tan^{-1}(-\gamma) = 2\tan^{-1}\left(-\frac{GM}{bv_\circ^2}\right), \tag{8.76}$$

a surprisingly compact result. The scattering angle $\phi_{\text{final}}$ will be negative, a consequence of the geometry of the setup.

Before proceeding further, imagine a situation where $M$ is extremely large, in which case $\gamma \to \infty$. Then $\phi_{\text{final}} = 2\tan^{-1}(-\infty) = 2(-\pi/2) = -\pi$: the probe will wrap around the planet and return toward the direction from whence it came! We will see, however, that is not a practical situation.

The position of closest approach can be determined by setting $dr/d\phi = 0$. This gives

$$\phi(r_{\min}) = \tan^{-1}(1/\gamma) = \tan^{-1}\left(\frac{bv_\circ^2}{GM}\right). \tag{8.77}$$

With this, you should be able to show that the distance of closest approach is

$$r_{\min} = \frac{b}{\gamma + \sqrt{1 + \gamma^2}}, \tag{8.78}$$

and that the speed at this position (the maximum speed) is

$$v_{\max} = v_\circ\left[\gamma + \sqrt{1 + \gamma^2}\right]. \tag{8.79}$$

To prove these it is helpful to know that for any angle $\theta$ you can write $\sin\theta = \tan\theta/\sqrt{1 + \tan^2\theta}$ and $\sin\theta = 1/\sqrt{1 + \tan^2\theta}$.

As a final (and slightly messy) calculation, we can establish an expression for estimating the time the probe will spend in the vicinity of the planet, the "fly-by" time. There is no rigorous definition of this, but this ambiguity will prove to be an advantage when choosing some limits of integration. Equation (8.72) gives the probe–planet distance as a function of $\phi$. If we imagine the probe covering some increment of angle $d\phi$, we can turn this into an increment of time via Equation (3.36):

$$dt = \frac{mr^2}{L}d\phi \;\Rightarrow\; dt = \frac{mb^2}{-mv_\circ b}\,\frac{d\phi}{[\gamma\,(1 + \cos\phi) + \sin\phi]^2}$$
$$= \frac{-b}{v_\circ}\,\frac{d\phi}{[\gamma\,(1 + \cos\phi) + \sin\phi]^2}.$$

The time to travel from $\phi = \phi_1$ to $\phi_2$ will be

$$t = \frac{-b}{v_\circ}\int_{\phi_1}^{\phi_2}\frac{d\phi}{[\gamma\,(1 + \cos\phi) + \sin\phi]^2}. \tag{8.80}$$

The ratio $b/v_\circ$ sets the timescale of the fly-by. Remarkably, there is a closed-form solution for integrals of this type:

$$\int\frac{dx}{[a\,(1 + \cos x) + c\sin x]^2}$$
$$= \frac{1}{c^3}\left[\frac{c\,(a\sin x - c\cos x)}{a\,(1 + \cos x) + c\sin x} - a\ln\left(a + c\tan\left(\frac{x}{2}\right)\right)\right]. \tag{8.81}$$

With the various sines and cosines here, it is convenient to take $\phi_1 = \pi/2$ and $\phi_2 = 0$, that is, we will get an expression for the time required for the probe to pass through 90° of trajectory from the "top" of the planet to the $x$-axis in Figure 8.10.

You might prefer different limits; this is the source of the ambiguity alluded to above. In Equation (8.81) we have $a = \gamma$ and $c = 1$. Flipping the limits of integration to eliminate the negative sign in (8.80) gives

$$\begin{aligned} t_{\text{flyby}} &= \frac{b}{v_\circ} \int_0^{\pi/2} \frac{d\phi}{[\gamma\,(1 + \cos\phi) + \sin\phi]^2} \\ &= \frac{b}{v_\circ}\left[\frac{(\gamma \sin\phi - \cos\phi)}{\gamma\,(1 + \cos\phi) + \sin\phi} - \gamma \ln\left(\gamma + \tan\left(\frac{\phi}{2}\right)\right)\right]_0^{\pi/2}. \end{aligned} \tag{8.82}$$

Evaluating the limits gives, after a bit of tidying-up,

$$t_{\text{flyby}} = \frac{b}{v_\circ}\left[\frac{2\gamma^2 + \gamma + 1}{2\gamma(\gamma + 1)} + \gamma \ln\left(\frac{\gamma}{\gamma + 1}\right)\right]. \tag{8.83}$$

As an example, consider a probe approaching Jupiter at $v_\circ = 30$ km s$^{-1}$. The mass of Jupiter is $1.898 \times 10^{27}$ kg, and its radius is 71,492 km (Table 5.1). Suppose we arrange for an impact parameter of three Jupiter radii, or $2.145 \times 10^8$ m. What will be the scattering angle? These numbers give

$$\gamma = \frac{GM}{bv_\circ^2} = \frac{(6.674 \times 10^{-11}\,\text{m}^3\,\text{kg}^{-1}\,\text{s}^{-2})(1.898 \times 10^{27}\,\text{kg})}{(2.145 \times 10^8\,\text{m})(30{,}000\,\text{ms}^{-1})^2} = 0.65616,$$

hence $\phi_{\text{final}} = 2\tan^{-1}(-0.65616) = -66°.54$, a quite significant trajectory alteration. The position of closest approach is $\phi(r_{\text{min}}) = +56°.73$. The distance of closest approach is about 115,800 km, about 1.6 Jupiter radii, and the speed at this point is about 55.6 km s$^{-1}$. The fly-by time is a mere 65 minutes. In such missions, the travel time to the planet may be years, only to have a close approach on the order of hours.

An interesting historical application of this result links to Einstein's general theory of relativity. Notice that the scattering angle $\phi_{\text{final}}$ is independent of the probe mass. We can then imagine applying Equation (8.76) to a *massless* particle, in particular a photon traveling at the speed of light $c = 2.998 \times 10^8$ m s$^{-1}$. Suppose that a photon just grazes the limb of the Sun, that is, that the impact parameter is equal to the radius of the Sun, $6.96 \times 10^8$ m. In this case we get

$$\gamma = \frac{GM}{bv_\circ^2} = \frac{(6.674 \times 10^{-11}\,\text{m}^3\,\text{kg}^{-1}\,\text{s}^{-2})(1.988 \times 10^{30}\,\text{kg})}{(6.96 \times 10^8\,\text{m})(2.998 \times 10^8\,\text{m s}^{-1})^2} = 2.1209 \times 10^{-6}.$$

The scattering effect is tiny, about $-2.43 \times 10^{-4}$ degrees, or $\sim -0.875$ seconds of arc. Curiously, Einstein's theory predicted the same effect, but with a magnitude exactly twice the Newtonian prediction worked out here.

How can such an effect be tested? Nature provided a way. Astronomers know that during a total eclipse of the Sun, background stars become visible. By photographing stars near the limb of the Sun during an eclipse and comparing their positions to angularly more distant (and hence largely unaffected) stars in an ordinary night-time photo of the same area of sky, the slight angular shifts could be observed and measured. This was first done during an eclipse in 1919, and the Einsteinian prediction proved to be correct. Today, astronomers routinely use space-based telescopes to secure distorted images of extremely distant galaxies whose light has been "gravitationally lensed" by closer-lying but still incredibly distant "foreground" galaxies. In effect, gravity now gives astronomers another wavelength band to add to their repertoire of optical, infrared, radio, microwave··· techniques. Another prediction of Einstein's theory, gravitational waves, has also been verified.

**Exercise:** In the Jupiter example above, what impact parameter is necessary to give $\phi_{\text{final}} = -40°$? How many Jupiter radii does this represent? Answer: About $3.87 \times 10^8$ m, or 5.4 Jupiter radii. This is a "softer" encounter than the above.

**Exercise:** You wish to arrange for a fly-by of Uranus in such a way that the distance of closest approach is one Uranus radius, so that the probe just skims the top of the atmosphere. If the distant speed of the probe is $v_\circ = 30$ km s$^{-1}$, what will be the necessary impact parameter? What will be the final scattering angle and fly-by time? You will need to solve for $\gamma$ numerically.

Answer: $\gamma \sim 0.205$; $b \sim 3.133 \times 10^7$ m $\sim 1.23$ Uranus radii; $\phi_{\text{final}} = -23°.2$; $t_{\text{flyby}} \sim 39$ minutes.

**Exercise:** To finish this book, here is a more challenging exercise. The analysis employed in this section should be applicable to an elliptical orbit, which would give an alternate approach to that of Sections 5.1 and 5.3. This approach is analogous to the latter section in that one assumes some of the characteristics of an elliptical orbit, but it is still legitimate. Here are some hints. Start with Equation (8.65), which is quite general for our gravitational force. Assume an attractor mass $M$ as usual, with $r$ and $\phi$ measured as usual. Start the orbiter (take $m \ll M$ for simplicity) at $(x, y) = (a(1 + \varepsilon), 0)$ (hence $\phi = 0$ initially) with speed $v_\circ$ in the $\hat{y}$ direction; these specify the initial linear and conserved angular momenta. This will change the limits in Equation (8.66), but the procedure is the same. When it comes to the energy analysis, don't forget the initial potential energy. Once you get a working expression for $r(\phi)$, impose one last condition: that when $\phi = \pi$, $r = a(1 - \varepsilon)$. You should be able to show that the standard equation for an ellipse emerges and that $L$ is given by Equation (5.10) with $\mu \rightarrow m$. Congratulations!

## 8.10 Some Final Words

We have come to the close of our whirlwind tour of orbital dynamics. We have covered much ground, but in some ways we have seen only the tip of a very large iceberg. For readers interested in further study, numerous related topics can be investigated: approximate methods of solving Kepler's equation; perturbations to orbits caused by the influence of other planets; central but non-inverse-square forces; unbound hyperbolic orbits such as those of extra-solar-system comets; analysis of binary-star data; and N-body analyses are but a few of the issues one can find addressed in more advanced texts. Some suggestions for further reading appear in Appendix C. I hope that this book has whetted your appetite to learn more of this fascinating and virtually bottomless topic, and thank you for coming this far.

## References

https://en.wikipedia.org/wiki/Lagrangepoint
https://en.wikipedia.org/wiki/UrbainLeVerrier
Reed, B. C. 2023, AmJPh, 91, in press
https://astro.unl.edu/classaction/animations/extrasolarplanets/ca_extrasolarplanets_starwobble.html
https://en.wikipedia.org/wiki/Barycenter
Edgar J. S. (ed) 2020, Observer's Handbook 2020 (23,; Royal Astronomical Society of Canada: Toronto)
Reed, B. C. 2014, EJPh, 35, 045009

# Appendix A

## Spherical Coordinates

This appendix gives a brief survey of spherical coordinates, without detailed derivations.

The system of spherical coordinates is illustrated in Figure A.1: $r$ is the radial distance from the origin to the point of interest ($0 < r < \infty$), $\theta$ is the "polar" angle measured from the positive-$z$-axis ($0 \leqslant \theta \leqslant \pi$), and $\phi$ is the "azimuthal" angle, measured counterclockwise from the positive-$x$-axis in the $xy$-plane ($0 \leqslant \phi \leqslant 2\pi$). Note that the $r$ coordinate here is different from that used for polar coordinates in Chapter 1. Also, "polar" is used here in a different sense than with the polar angle $\phi$ of $xy$-plane polar coordinates. Spherical coordinates will be needed if you are considering, say, orbits inclined to the plane of the ecliptic, transfer orbits involving out-of-plane motion, or analyzing stellar systems in mutual orbits inclined to the plane of the sky. Be aware that some texts swap the definitions of $\theta$ and $\phi$, although the definitions used here are common in the physics community.

The transformation equations from Cartesian to spherical coordinates are

$$\begin{aligned} r &= \sqrt{x^2 + y^2 + z^2} \\ \theta &= \cos^{-1}(z/r) \\ \phi &= \tan^{-1}(y/x), \end{aligned} \tag{A.1}$$

and the reverse transformations from spherical to Cartesian are

$$\begin{aligned} x &= r \sin\theta \, \cos\phi \\ y &= r \sin\theta \, \sin\phi \\ z &= r \cos\theta. \end{aligned} \tag{A.2}$$

In terms of the Cartesian unit vectors, the spherical coordinate unit vectors are

$$\begin{aligned} \hat{\boldsymbol{r}} &= (\sin\theta\cos\phi)\,\hat{\boldsymbol{x}} + (\sin\theta\sin\phi)\,\hat{\boldsymbol{y}} + (\cos\theta)\,\hat{\boldsymbol{z}} \\ \hat{\boldsymbol{\theta}} &= (\cos\theta\cos\phi)\,\hat{\boldsymbol{x}} + (\cos\theta\sin\phi)\,\hat{\boldsymbol{y}} - (\sin\theta)\,\hat{\boldsymbol{z}} \\ \hat{\boldsymbol{\phi}} &= -(\sin\phi)\,\hat{\boldsymbol{x}} + (\cos\phi)\,\hat{\boldsymbol{y}}. \end{aligned} \tag{A.3}$$

doi:10.1088/2514-3433/acb430ch9

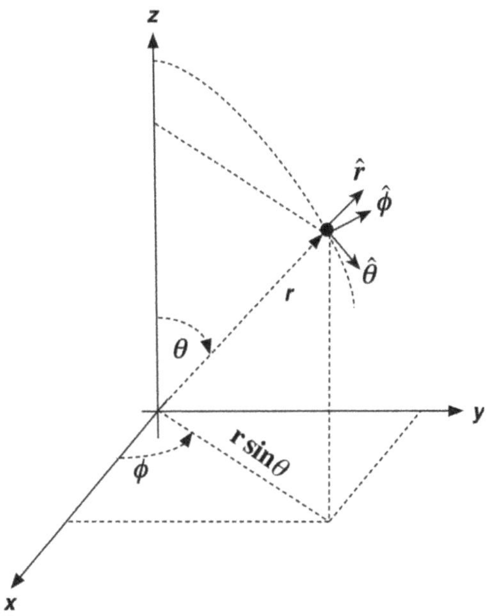


**Figure A.1.** Spherical coordinates $(r, \theta, \phi)$. Be aware that some texts swap the definitions of $\theta$ and $\phi$. The arrows attached to the point of interest (the black dot) show the directions of the spherical coordinate unit vectors at that point.

The Cartesian unit vectors in terms of the spherical coordinate unit vectors are

$$\begin{aligned} \hat{x} &= (\sin\theta\cos\phi)\,\hat{r} + (\cos\theta\cos\phi)\,\hat{\theta} - (\sin\phi)\,\hat{\phi} \\ \hat{y} &= (\sin\theta\sin\phi)\,\hat{r} + (\cos\theta\sin\phi)\,\hat{\theta} + (\cos\phi)\,\hat{\phi} \\ \hat{z} &= (\cos\theta)\,\hat{r} - (\sin\theta)\,\hat{\theta}. \end{aligned} \tag{A.4}$$

As with polar coordinates, Cartesian unit vectors are considered to be fixed in space and act like constants, while spherical coordinate unit vectors are functions of the direction $(\theta, \phi)$ of the point under consideration. The spherical coordinate unit vectors are illustrated in Figure A.1.

Like the Cartesian and polar systems, spherical coordinates form a right-handed system, with the same rules for forming scalar ("dot") and vector ("cross") products. Figure A.2 shows pictorial triads as way of remembering this. Suppose that you wish to compute $\hat{\theta} \times \hat{r}$. Start by locating $\hat{\theta}$ in the triad. Follow the circle until you get to $\hat{r}$ by the shortest route; this requires a clockwise motion. Keep going clockwise until you hit the next unit vector in the triad, which is $\hat{\phi}$. But a clockwise cross-product always gives a negative answer (try it with your right hand—your thumb will naturally point "down"), so $\hat{\theta} \times \hat{r} = -\hat{\phi}$. You should be able to prove this as well by crossing $\hat{\theta}$ and $\hat{r}$ from Equations (A.3). Also as with Cartesian unit vectors, any spherical unit vector dotted into itself gives unity, and crossed into itself gives zero.

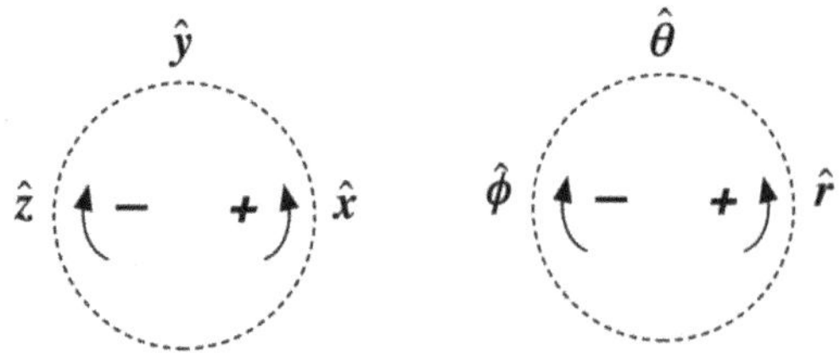


**Figure A.2.** Mnemonic triads for remembering the results of spherical coordinate unit-vector cross-products. If a counterclockwise movement is required to bring the first unit vector in a cross-product into the second one, the result is the positive of the next unit vector found by proceeding in the same direction. A clockwise movement gives the negative of the next unit vector found by proceeding in the same direction.

The time-derivatives of the spherical unit vectors are

$$\frac{d\hat{\boldsymbol{r}}}{dt} = (\dot{\theta})\,\hat{\theta} + (\phi \sin\theta)\hat{\boldsymbol{\phi}}, \tag{A.5}$$

$$\frac{d\hat{\theta}}{dt} = -(\dot{\theta})\,\hat{\boldsymbol{r}} + (\dot{\phi}\cos\theta)\hat{\boldsymbol{\phi}}, \tag{A.6}$$

and

$$\frac{d\hat{\boldsymbol{\phi}}}{dt} = -(\dot{\phi}\sin\theta)\,\hat{\boldsymbol{r}} - (\dot{\phi}\cos\theta)\hat{\theta}. \tag{A.7}$$

The position, velocity, acceleration, angular momentum, and torque vectors are

$$\boldsymbol{r} = r\hat{\boldsymbol{r}}, \tag{A.8}$$

$$\boldsymbol{v} = \dot{r}\,\hat{\boldsymbol{r}} + (r\,\dot{\theta})\,\hat{\boldsymbol{\theta}} + (r\,\dot{\phi}\sin\theta)\,\hat{\boldsymbol{\phi}}, \tag{A.9}$$

$$\begin{aligned}\boldsymbol{a} = &\left\{\ddot{r} - r(\dot{\theta})^2 - r(\dot{\phi})^2\sin^2\theta\right\}\hat{\boldsymbol{r}} + \left\{2\dot{r}\,\dot{\theta} + r\,\ddot{\theta} - r\,(\dot{\phi})^2\sin\theta\cos\theta\right\}\hat{\boldsymbol{\theta}}\\ &+\{2\dot{r}\,\dot{\phi}\sin\theta + 2r\,\dot{\theta}\,\dot{\phi}\cos\theta + r\,\ddot{\phi}\sin\theta\}\hat{\boldsymbol{\phi}},\end{aligned} \tag{A.10}$$

$$\boldsymbol{L} = -(mr^2\,\dot{\phi}\sin\theta)\hat{\boldsymbol{\theta}} + (m\,r^2\,\dot{\theta})\hat{\boldsymbol{\phi}}, \tag{A.11}$$

and

$$\begin{aligned}\boldsymbol{\tau} = &-m\{2r\,\dot{r}\,\dot{\phi}\sin\theta + r^2\ddot{\phi}\sin\theta + 2r^2\,\dot{\theta}\,\dot{\phi}\cos\theta\}\hat{\boldsymbol{\theta}}\\ &+m\{r^2\ddot{\theta} + 2r\,\dot{r}\,\dot{\theta} - r^2(\dot{\phi})^2\sin\theta\cos\theta\}\hat{\boldsymbol{\phi}}.\end{aligned} \tag{A.12}$$

The total energy is

$$E = \frac{1}{2}m[(\dot{r})^2 + (r\,\dot{\theta})^2 + (r\,\dot{\phi}\sin\theta)^2] + V. \tag{A.13}$$

The gradient operator for a scalar function $V(r, \theta, \phi)$ is

$$\nabla V = \left(\frac{\partial V}{\partial r}\right)\hat{\boldsymbol{r}} + \left(\frac{1}{r}\frac{\partial V}{\partial \theta}\right)\hat{\boldsymbol{\theta}} + \left(\frac{1}{r\sin\theta}\frac{\partial V}{\partial \phi}\right)\hat{\boldsymbol{\phi}}. \tag{A.14}$$

The curl of a vector function $\boldsymbol{F}$ is

$$\begin{aligned}\nabla \times \boldsymbol{F} = {} & \frac{1}{r\sin\theta}\left[\frac{\partial}{\partial \theta}(\sin\theta F_\phi) - \frac{\partial F_\theta}{\partial \phi}\right]\hat{\boldsymbol{r}} + \frac{1}{r}\left[\frac{1}{\sin\theta}\frac{\partial F_r}{\partial \phi} - \frac{\partial}{\partial r}(r\,F_\phi)\right] \\ & \hat{\boldsymbol{\theta}} \\ & + \frac{1}{r}\left[\frac{\partial}{\partial r}(rF_\theta) - \frac{\partial F_r}{\partial \theta}\right]\hat{\boldsymbol{\phi}},\end{aligned} \tag{A.15}$$

where $(F_r, F_\theta, F_\phi)$ are the components of $\boldsymbol{F}$ in the form

$$\boldsymbol{F} = F_r\hat{\boldsymbol{r}} + F_\theta\hat{\boldsymbol{\theta}} + F_\phi\hat{\boldsymbol{\phi}}. \tag{A.16}$$

# Appendix B

## Circular-Orbit Perturbation Theory for Non-Inverse-Square Central Forces

This section parallels the development of circular orbit radial perturbation theory in Section 5.10 for potentials of the form $V(r) = -Kr^n$, the same form as used in the analysis of the shell-point equivalency theorem in Section 3.7. The force corresponding to this potential is $\boldsymbol{F} = -(\partial V/\partial r)\hat{\boldsymbol{r}} = +Knr^{n-1}\hat{\boldsymbol{r}}$. If $n$ is negative, $K$ must be positive to ensure an attractive force; similarly, $K$ must be negative if $n$ is positive to give an attractive force.

With this potential, the effective potential $U(r)$ of (5.75) is

$$U(r) = -Kr^n + \frac{h}{r^2}, \tag{B.1}$$

where

$$h = \frac{L^2}{2m}. \tag{B.2}$$

The minimum of $U(r)$ (the circular orbit radius) occurs at

$$r_o = \left(-\frac{2h}{nK}\right)^{\frac{1}{n+2}}, \tag{B.3}$$

at which position the effective potential has the value

$$U(r_o) = \frac{h}{r_o^2}\left[\frac{2}{n} + 1\right]. \tag{B.4}$$

From a development like that in Section 5.8, for a mass $m$ in a circular orbit of radius $r$, Kepler's third law gives the square of the period as

$$T_o^2 = \frac{4\pi^2 m}{|nK|} r^{2-n}. \tag{B.5}$$

doi:10.1088/2514-3433/acb430ch10

We will return to this expression presently.

Now define a new radial coordinate $x$ as

$$x = r - r_0. \tag{B.6}$$

As in Section 5.10, $x$ is *not* the usual Cartesian coordinate; rather, it is defined to enable us to shift the origin of the radial coordinate so that $x = 0$ corresponds to the position of the minimum of the $U(r)$ curve. In terms of $x$,

$$\begin{aligned} U(x) &= -K(x + r_0)^n + \frac{h}{(x + r_0)^2} \\ &= -Kr_o^n(1 + x/r_0)^n + \frac{h}{r_0^2(1 + x/r_0)^2}, \end{aligned} \tag{B.7}$$

where the last step follows on factoring out $r_0$ from within each bracket.

Now, as with the inverse-square case, suppose that the orbiter receives a slight *purely radial* push inwards or outwards. The displacement $x$ is again presumed to be small compared to $r_0$, and we will perform binomial expansions on both terms in Equation (B.7); see Equation (1.23). On doing so and gathering terms in the same power of $x$, the result is

$$\begin{aligned} U(x) \sim & \left[-Kr_o^n + \frac{h}{r_0^2}\right] + \left[-nKr_o^n - 2\frac{h}{r_0^2}\right]\left(\frac{x}{r_0}\right) \\ &+ \left[-Kr_o^n\frac{n(n-1)}{2} + 3\frac{h}{r_0^2}\right]\left(\frac{x}{r_0}\right)^2 + \cdots. \end{aligned} \tag{B.8}$$

As an exercise, check that if $K = GMm$ and $n = -1$, you reproduce the results of Section 5.10; note that $K$ is the negative of our usual $\kappa = -GMm$.

If you substitute for $r_0$ from Equation (B.3), you will find that, as before, the first term on the right side of (B.8) is just $U(r_0)$ of Equation (B.4), and that the middle term vanishes. In analogy to (5.90), what remains is, after some simplification,

$$U(x) \sim U(r_0) + \frac{1}{2}\left[\frac{2h(n+2)}{r_0^4}\right]x^2 + \cdots. \tag{B.9}$$

The term in square brackets is our new effective spring constant. As in Section 5.10, the energy analysis then gives Equation (5.95) again:

$$m\left(\frac{d^2x}{dt^2}\right) = -kx. \tag{B.10}$$

*Here is the important physical point. To have a restoring force, that is, one where the force is negative for $x > 0$ and positive for $x < 0$, we must have $k > 0$. This can only happen if $n > -2$ in Equation (B.9). So, as advertised in Section 5.10, circular orbits in $-Kr^n$ potentials will be stable against radial perturbations for $n > -2$.*

The square of the period for radial oscillations in response to a perturbation is

$$T_{osc}^2 = 4\pi^2\left(\frac{m}{k}\right) = 4\pi^2\, m\left[\frac{r_o^4}{2h(n+2)}\right] = \frac{4\pi^2 m^2 r_o^4}{(n+2)L^2}, \tag{B.11}$$

where Equation (B.2) was used to eliminate $h$.

As before, we compare this to the orbital period to see if a perturbation will result in a closed orbit or not. For this we need Equation (B.5), using $r_o$ for $r$:

$$T_o^2 = \frac{4\pi^2 m}{|nK|} r_o^{2-n}. \tag{B.12}$$

To compare $T_{osc}$ and $T_o$, we need to get $L^2$ in terms of $|nK|$, or the other way around. I will convert $|nK|$ to $L$.

For our circular-orbiting mass in this potential, a centripetal force analysis tells us that

$$|nK| r_o^{n-1} = \frac{mv^2}{r_o}. \tag{B.13}$$

Now, for a circular orbit, $L = mvr_o$. Using this to eliminate $v$ as $v = L/mr_o$ gives

$$|nK| = \frac{L^2}{mr_o^{n+2}}, \tag{B.14}$$

which upon substitution into Equation (B.12) gives

$$T_o^2 = \frac{4\pi^2 m^2 r_o^4}{L^2}. \tag{B.15}$$

Compare Equations (B.11) and (B.15):

$$\frac{T_{osc}^2}{T_o^2} = \frac{1}{n+2}. \tag{B.16}$$

Now, we must have $n > -2$ for a stable orbit, so $T_{osc} < T_o$: for permissible *integer* values of $n = -1, 0, 1, 2, \ldots$, the radial oscillations must have periods less than or at most equal to that of the circular orbit. If the orbital period is then an exact integer multiple, say $N$, of the radial period, that is, if $T_o = NT_{osc}$, then the orbit will be closed even in the presence of the radial perturbation: After $N$ radial-oscillation periods, the orbiter will return to the position it would have had if the perturbation had not happened. This requires

$$\sqrt{n+2} = N. \tag{B.17}$$

This is satisfied for $n = -1$ (inverse-square force), $n = 2$ (harmonic oscillator force), as well as $n = 7$, 14, and so forth.

Rational-fraction values of $n$, although apparently not preferred in Nature, can also lead to closed orbits. For example, suppose that the period for five radial oscillations is exactly equal to that for two orbits, $T_{osc} = (2/5)T_o$. Then you will find that $n = 17/4$. In general, if an integer number if radial oscillations "fit into" some integer number of circular orbits, $n$ will be a rational fraction.

# Appendix C

## Further Reading

### Journal Articles

Three journals which are aimed at undergraduate-level students and their instructors are *American Journal of Physics*, *European Journal of Physics*, and *The Physics Teacher*. These well-regarded publications frequently carry papers on Kepler's laws and orbital mechanics, and should be available in most college and university libraries. The following list offers a sampling of papers appearing in these journals from over the last few years; further references can be found within these. The websites of these journals can be searched for topics of interest.

Bates A 2013, Galilean moons, Kepler's third law, and the mass of Jupiter, *Phys. Teach.* **51** 428–429. See correction in *Phys. Teach.* **51** 212–214

Benacka J 2014, On planetary motion—a way to solve the problem and a spreadsheet simulation, *Eur. J. Phys.* **35**(4) 045016

Bracco C and Provost J-P 2009, Had the planet Mars not existed: Kepler's equant model and its physical consequences, *Eur. J. Phys.* **30**(5) 1085–1093

Bucher M 1998, Kepler's third law: Equal volumes in equal times, *Phys. Teach.* **36**(4) 212–214

Carroll B W 2019, The delicate dance of orbital rendezvous, *Am. J. Phys.* **87**(8) 627–637

Davies B 1983, Elementary theory of perihelion precession, *Am. J. Phys.* **51**(10) 909–911

Easton R W, Anderson R L and Lo M W 2022, Conic transfer arcs for Kepler problem, *Am. J. Phys.* **90**(9) 666–671

Edlund E M 2021, Interception and rendezvous: An intuition-building approach to orbital dynamics, *Am. J. Phys.* **89**(6) 559–566

Gatland I 2022, Gravitational orbits and the Lambert problem, *Am. J. Phys.* **90**(3) 177–178

Hecht E 2021, The true story of Newtonian gravity, *Am. J. Phys.* **89**(7) 683–692

doi:10.1088/2514-3433/acb430ch11

Hsiang W Y, Chang H C, Yao H and Chen P J 2011, An alternative way to achieve Kepler's laws of equal areas and ellipses for the Earth, *Eur. J. Phys.* **32**(5) 1405–1412

Hsiang W Y, Chang H C, Yao H and Lee P S 2015, Re-establishing Kepler's first two laws for planets in a concise way through the non-stationary Earth, *Eur. J. Phys.* **36**(4) 045006

Newton I and Henry R C 2000, Circular motion, *Am. J. Phys.* **68**(7) 637–639

Oostra B 2015, Introducing Earth's orbital eccentricity, *Phys. Teach.* **53** 554–556.

Provost J-P and Bracco C 2009, A simple derivation of Kepler's laws without solving differential equations, *Eur. J. Phys.* **30**(3) 581–586

Rovšek B 2021, How eccentric is the orbit of the Earth, and where is the Sun?, *Phys. Teach.* **59** 438–439

Simha A 2021, An algebra and trigonometry-based proof of Kepler's first law, *Am. J. Phys.* **89**(11) 1009–1011

Vogt E 1996, Elementary derivation of Kepler's laws, *Am. J. Phys.* **64**(4) 392–396

## General Works and Texts

Probably no aspect of classical physics has been more studied than celestial mechanics. The sources listed here are but a tiny fraction of what is available. I list these because they give excellent semipopular accounts of the discoveries, and biographies of the people involved. Richard Westfall's exhaustive biography of Newton is regarded as the standard in the field, and Thornton and Marion is a well-regarded undergraduate-level physics text. Newton's Principia is available in published and online versions, but is of course heavy going.

Connor J A 2004 *Kepler's Witch: An Astronomer's Discovery of Cosmic Order Amid Religious War, Political Intrigue, and the Heresy Trial of His Mother* (New York: Harper Collins)

Curtis H D 2020 *Orbital Mechanics for Engineering Students*, 4th edn (Oxford: Butterworth-Heinemann)

Ferguson K 2002 *Tycho and Kepler: The Unlikely Partnership that Forever Changed Our Understanding of the Heavens* (New York: Walker & Company)

Koestler A 1959 *The Sleepwalkers. A History of Man's Changing Vision of the Universe* (New York: Macmillan)

Newton I 1687 *Philosophiae Naturalis Principia Mathematica* (London: Royal Society). How can a book on orbital mechanics not reference Newton? An online copy of the original in Latin can be found at http://www.gutenberg.org/ebooks/28233, and an English translation at https://openlibrary.org/books/OL7089085M/Newton

Sobel D 2011 *A More Perfect Heaven: How Copernicus Revolutionized the Cosmos* (New York: Walker & Company)

Sobel D 2005 *The Planets* (New York: Viking)

Thornton S T and Marion J B 2008 *Classical Dynamics of Particles and Systems* 5th edn (Boston, MA: Brooks/Cole CENGAGE Learning)

Westfall R S 1980 *Never at Rest: A Biography of Isaac Newton* (Cambridge: Cambridge University Press)

# Appendix D

## Summary of Useful Formulae

**Polar Coordinates**

$$r = \sqrt{x^2 + y^2} \qquad x = r \cos\phi$$
$$\phi = \tan^{-1}(y/x) \qquad y = r \sin\phi$$

$$\hat{\boldsymbol{r}} = (\cos\phi)\hat{\boldsymbol{x}} + (\sin\phi)\hat{\boldsymbol{y}} \qquad \hat{\boldsymbol{x}} = (\cos\phi)\,\hat{\boldsymbol{r}} - (\sin\phi)\,\hat{\boldsymbol{\phi}}$$
$$\hat{\boldsymbol{\phi}} = -(\sin\phi)\hat{\boldsymbol{x}} + (\cos\phi)\hat{\boldsymbol{y}} \qquad \hat{\boldsymbol{y}} = (\sin\phi)\,\hat{\boldsymbol{r}} + (\cos\phi)\,\hat{\boldsymbol{\phi}}$$

$$\frac{d\,\hat{\boldsymbol{r}}}{dt} = (\dot{\phi})\,\hat{\boldsymbol{\phi}} \qquad \frac{d\hat{\boldsymbol{\phi}}}{dt} = -(\dot{\phi})\,\hat{\boldsymbol{r}}$$

**Central-Force Dynamics**

$$\boldsymbol{r} = r\,\hat{\boldsymbol{r}} \qquad \boldsymbol{v} = \dot{r}\,\hat{\boldsymbol{r}} + (r\,\dot{\phi}\,)\,\hat{\boldsymbol{\phi}}$$

$$\boldsymbol{a} = \{\ddot{r} - r(\dot{\phi})^2\}\hat{\boldsymbol{r}} + \{2\dot{r}\,\dot{\phi} + r\,\ddot{\phi}\}\hat{\boldsymbol{\phi}}.$$

$$\boldsymbol{L} = (mr^2\,\dot{\phi})\hat{\boldsymbol{z}} \qquad \boldsymbol{\tau} = m\{2r\,\dot{r}\,\dot{\phi} + r^2\ddot{\phi}\}\hat{\boldsymbol{z}}$$

$$E = \frac{1}{2}m[(\dot{r})^2 + (r\,\dot{\phi})^2] + V$$

$$\mu = \left(\frac{M\,m}{M + m}\right).$$

$$\boldsymbol{F} = -\frac{GMm}{r^2}\hat{\boldsymbol{r}} = -\nabla V = -\left(\frac{\partial V}{\partial r}\right)\hat{\boldsymbol{r}}$$

$$V(r) = -\frac{GMm}{r} = +\frac{\kappa}{r}$$

$$d\phi = \left(\frac{L}{\mu r^2}\right)dt \iff dt = \left(\frac{\mu r^2}{L}\right)d\phi$$

doi:10.1088/2514-3433/acb430ch12

$$E = \frac{1}{2}\left[\mu\left(\frac{dr}{dt}\right)^2 + \frac{L^2}{\mu r^2}\right] + V(r)$$

$$t - t_o = \int_{r_O}^{r} \frac{dr}{\sqrt{(2/\mu)[E - V(r) - L^2/2\mu r^2]}}$$

$$\phi - \phi_o = \int_{r_O}^{r} \frac{(L/\mu r^2)}{\sqrt{\frac{2}{\mu}[E - V(r) - L^2/2\mu r^2]}} dr$$

$$F(r) = -\left(\frac{L^2}{\mu r^2}\right)\left\{\frac{1}{r^2}\left(\frac{d^2 r}{d\phi^2}\right) - \frac{2}{r^3}\left(\frac{dr}{d\phi}\right)^2 - \frac{1}{r}\right\}$$

$$F(u) = +\frac{L^2 u^2}{\mu}\left\{\frac{d^2 u}{d\phi^2} + u\right\}$$

**Ellipse Geometry (See Figure** D.1)

$$r = \frac{a(1 - \varepsilon^2)}{1 - \varepsilon \cos\varphi} = a(1 + \varepsilon \cos\psi)$$

$$\tan\left(\frac{\phi}{2}\right) = \sqrt{\frac{(1 - \varepsilon)}{(1 + \varepsilon)}} \tan\left(\frac{\psi}{2}\right)$$

$$A_{\text{ellipse}} = \pi\ a\ b = \pi\ a^2\sqrt{1 - \varepsilon^2}$$

**Elliptical Orbit Dynamics**

$$L^2 = GMm\mu a(1 - \varepsilon^2) = \frac{G(Mm)^2}{(M + m)} a(1 - \varepsilon^2)$$

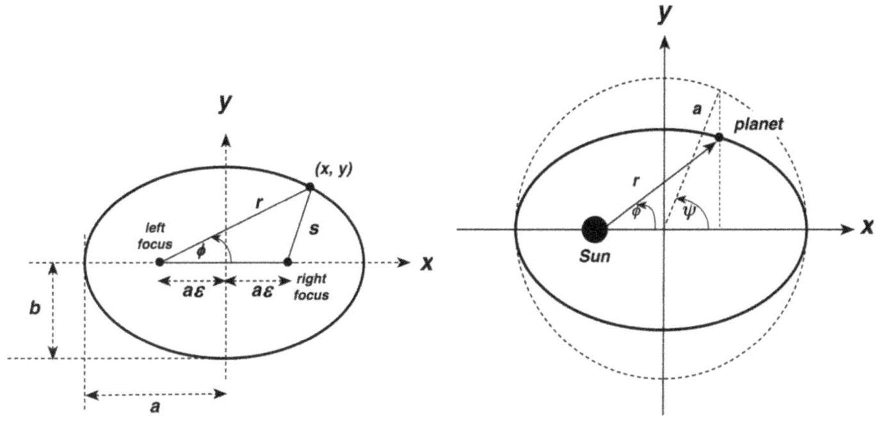


**Figure D.1.** Ellipse geometries.

$$L = \frac{GT(Mm)}{2\pi a}\sqrt{1-\varepsilon^2} = \frac{2\mu\pi a^2}{T}\sqrt{1-\varepsilon^2}.$$

$$v^2 = \frac{G(M+m)}{a(1-\varepsilon^2)}(1+\varepsilon^2 - 2\varepsilon\cos\phi) = G(M+m)\left(\frac{2}{r} - \frac{1}{a}\right)$$

$$E = -\frac{GMm}{2a}$$

$$T^2 = \left[\frac{4\pi^2}{G(M+m)}\right]a^3$$

$$\left(\frac{t}{T}\right)_{0\to\phi} = \frac{\sqrt{1-\varepsilon^2}}{2\pi}\left[\frac{\varepsilon\sin\phi}{(1-\varepsilon\cos\phi)} + \frac{2}{\sqrt{1-\varepsilon^2}}\tan^{-1}\left[\sqrt{\frac{1+\varepsilon}{1-\varepsilon}}\tan(\phi/2).\right]\right]$$

$$\omega t_{0\to\psi} = \psi + \varepsilon\sin\psi$$

$$v_{\text{esc}} = \sqrt{\frac{2GM}{R}}$$

**Average Distance and Speed**

$$\langle r\rangle_t = a(1+\varepsilon^2/2) \qquad \langle r\rangle_\phi = a\sqrt{1-\varepsilon^2} \qquad \langle r\rangle_S = a$$

$$\langle v\rangle_t = \frac{2\pi a}{T}\left(1 - \frac{1}{4}\varepsilon^2 - \frac{3}{64}\varepsilon^4 - \frac{5}{256}\varepsilon^6 + \cdots\right)$$

$$\langle v^2\rangle_t = \frac{G(M+m)}{a}$$

**Gravitational Scattering**

$$\gamma = \frac{GM}{bv_o^2} \qquad r = \frac{b}{(\gamma + \sin\phi + \gamma\cos\phi)}$$

$$\phi_{\text{final}} = 2\tan^{-1}(-\gamma) = 2\tan^{-1}\left(-\frac{GM}{bv_o^2}\right)$$

$$\phi(r_{\min}) = \tan^{-1}(1/\gamma) = \tan^{-1}\left(\frac{bv_o^2}{GM}\right)$$

$$r_{\min} = \frac{b}{\gamma + \sqrt{1+\gamma^2}} \qquad v_{\max} = v_o\left[\gamma + \sqrt{1+\gamma^2}\right]$$

$$t_{\text{flyby}} = \frac{b}{v_o}\left[\frac{2\gamma^2+\gamma+1}{2\gamma(\gamma+1)} + \gamma\ln\left(\frac{\gamma}{\gamma+1}\right)\right]$$

# Appendix E

## Glossary of Symbols

Some symbols are used to represent more than one quantity. In some cases, section numbers where a symbol is introduced or prominently used are indicated in square brackets.

**Table E.1.** Glossary of Symbols.

| Symbol | Meaning and Comments |
|---|---|
| $\boldsymbol{A}$ | Laplace–Runge–Lenz vector [5.7] |
| $\boldsymbol{a}$ | Acceleration vector [2.1, 3.6] |
| $a$ | Semimajor axis of elliptical orbit [4.1]; also the magnitude of the acceleration vector Occasionally used as a general parameter in an integral |
| $b$ | Semiminor axis of elliptical orbit [4.1]. Occasionally used as a parameter in an integral; also impact parameter in gravitational scattering [8.9] |
| $E$ | Total energy of two-mass system [2.1, 5.2] |
| $\boldsymbol{F}$ | Force vector; magnitude $F$ [3.2] |
| $G$ | Newtonian gravitational constant; $G = 6.674 \times 10^{-11}$ m$^3$ kg$^{-1}$ s$^{-2}$ |
| $g$ | Acceleration due to gravity at Earth's surface, 9.8 m s$^{-2}$; occasionally a general constant, notably in perturbation theory of Section 5.10 and Appendix B |
| $h$ | Constant in perturbation theory of Section 5.10 and Appendix B |
| $K$ | Kinetic energy of orbiting "reduced mass" [3.2] |
| $k$ | Factor in analysis of perihelion precession [8.5]; occasionally a general constant, "spring constant" in analysis of perturbation theory [5.10] |
| $\boldsymbol{L}$ | Angular momentum of two-body system; magnitude $L$ [2.1, 3.5, 5.1] |
| $M$ | Mass; generally the larger of the two masses in a two-mass system |
| $m$ | Mass; generally the smaller of the two masses in a two-mass system |
| $M_E$ | Mass of Earth; $5.972 \times 10^{24}$ kg |
| $Q$ | Parameter in analysis of orbital angular position-radial distance analysis [5.3] |
| $R_E$ | Radius of Earth (6371 km); radius of Earth's orbit in analysis of Hohmann transfer [7.1] |

(*Continued*)

doi:10.1088/2514-3433/acb430ch13

**Table E.1.** *(Continued)*

| Symbol | Meaning and Comments |
|---|---|
| $R_{\text{Schwarz}}$ | Schwarzschild radius for black hole [5.11] |
| $r$ | Polar radial coordinate [1.1, 4.1]; general radial distance from focus in elliptical orbit |
| $r_0$ | Position of minimum effective potential $U(r)$; [5.9], [5.10], and Appendix B |
| $\langle r\rangle_t$ | Time-averaged distance of planet from force center [8.1.2] |
| $\langle r\rangle_\phi$ | Angle-averaged distance of planet from force center [8.1.3] |
| $\langle r\rangle_S$ | Arc length-averaged distance of planet from force center [8.1.4] |
| $s$ | empty-focus distance to orbit in ellipse [4.1]; Alice's lead distance in ham-sandwich toss [7.3] |
| $S$ | perimeter of ellipse [8.1.4] |
| $T$ | Orbital period |
| $t$ | Time as a variable |
| $U(r)$ | Effective potential [5.9] |
| $u$ | Proxy radial coordinate; $u = 1/r$ [3.7] |
| $V$ | Potential energy [2.1] and [3.2]; usually as a potential energy function $V(r)$ |
| $v$ | Speed; magnitude of velocity vector $\boldsymbol{v}$ |
| $v_{\text{esc}}$ | Escape velocity [5.11] |
| $\langle v\rangle_t$ | Time-averaged orbital speed [8.2] |
| $W$ | Work [3.2] |
| $\beta$ | Parameter in ellipse geometry [5.1]; parameter in calculation of time an orbiting object spends within distance $r$ of the force center [Chapter 6] |
| $\delta$ | Factor in analysis of perihelion precession [8.5] |
| $\varepsilon$ | Orbital eccentricity |
| $\gamma$ | Dimensionless parameter in analysis of gravitational scattering [8.9] |
| $\kappa$ | Parameter in gravitational analysis; $\kappa = -GMm$ [3.2] |
| $\mu$ | Reduced mass; $\mu = Mm/(M + m)$ |
| $\phi$ | Polar coordinate azimuthal angle [1.1]; general orbital apsidal angle; true anomaly |
| $\psi$ | Eccentric anomaly in Kepler's equation [Chapter 6]; angle in area of triangle [4.3] |
| $\tau$ | Torque [2.1] |
| $\theta$ | Spherical coordinate polar angle [Appendix A] |
| $\omega$ | Angular speed of orbiting object in analysis of Kepler's equation [Chapter 6] |

CPSIA information can be obtained
at www.ICGtesting.com
Printed in the USA
BVHW010723030423
661609BV00002B/14

9 780750 356060